THE MACHINE THAT MOVES ELECTRICITY

The Hidden Technology That Built the Grid – and Could Determine America's Energy Future

Douglas Donovan

ISBN: 978-1-0881-5285-0

Companion volumes by the same author: A History of the United States Electrical Grid; The Reliability Frontier: How America's Grid Is Racing to Meet the AI Era.

Preface

There is a piece of equipment standing in a fenced gravel lot within a mile or two of almost every reader of this book, and most of those readers have never once looked at it. It has no screen, no blinking lights, no brand name stamped anywhere a passerby would notice. It hums, faintly, on humid days. It is usually gray or dull green, roughly the size of a delivery van, and it is, without exaggeration, one of the most important machines human beings have ever built. It is a power transformer, and it is the reason the electricity generated at a plant hundreds of miles away can arrive at your kitchen outlet at exactly the voltage your toaster oven expects.

I wrote two earlier books before this one without giving the transformer the attention it deserved. *A History of the United States Electrical Grid* told the story of how America's electrical system grew from Edison's Pearl Street station into the vast interconnected machine that powers 140 million customers today, and it mentioned transformers on nearly every page without ever stopping to explain what one actually is. *The Reliability Frontier* described how artificial intelligence datacenters are straining that same grid to its limits in the mid-2020s, and it kept circling back to a single, frustrating fact: utilities cannot get the equipment they need to keep up, because the factories that build large power transformers cannot build them fast enough. In both books, the transformer sat quietly in the background, doing indispensable work that the story never paused to examine.

This book is that pause. It is the story of a single category of machine, told from Michael Faraday's laboratory in 1831 to the transformer factories racing to expand capacity in 2026, and it argues that this unglamorous piece of iron, copper, and oil is now one of the most consequential objects in the American economy. When you read that a utility has told a hyperscale datacenter developer it will take four years to get power, the sentence hiding inside that delay is almost always: we are waiting for a transformer.

I have written this book, as I wrote the previous two, for the general reader and the engaged undergraduate – someone who has heard the word "transformer" used in a news article about grid strain or AI energy demand and wants to understand, from first principles, what the device actually does, why it was invented, and why it has suddenly become so hard to buy one. Where technical concepts are unavoidable – electromagnetic induction, core losses, extra-high-voltage insulation – I have tried to build the vocabulary a general reader needs as we go, the same way I approached capacity markets and grid-forming inverters in *The Reliability Frontier*.

This book can be read entirely on its own. It does not require having read either of its companion volumes, and it explains every concept it needs from scratch. But readers of *A History of the United States Electrical Grid* will recognize familiar landmarks – Niagara Falls, rural electrification, the deregulation era – seen this time through the lens of the machine that made each of them possible. And readers of *The Reliability Frontier* will find, in the book's final chapters, the engineering and industrial explanation for a crisis that book could only describe from the outside.

The transformer has been sitting in that gravel lot, unremarked, for well over a century. It is time someone looked at it closely.

Introduction: The Machine You've Never Noticed

Picture the electricity in your home tracing its path backward, like a river flowing in reverse. It leaves your outlet, travels through the wiring in your walls, and reaches a small gray canister mounted on a utility pole or squatting in a green box on your lawn. That canister – about the size of a large trash can – is a distribution transformer, and it has just taken electricity arriving at somewhere between 4,000 and 13,000 volts and reduced it to the 120 or 240 volts your appliances actually use. Continue backward past that transformer, along the line that feeds it, and within a few miles you will reach a substation: a fenced yard filled with larger transformers, each the size of a small building, stepping voltage down from the tens of thousands of volts used for regional distribution. Continue further, along a high-voltage transmission line strung between steel towers, and you will eventually reach a switchyard beside a power plant, where the largest transformers in the entire system – machines that can weigh four hundred tons and cost several million dollars each – take the roughly 20,000 volts produced at the generator and step it up to 230,000, 345,000, 500,000, or even 765,000 volts for the journey across the countryside.

At every one of those points, nothing about the electricity itself has changed in kind. The same alternating current, oscillating sixty times a second, flows through the entire chain from generator to toaster. What has changed, repeatedly and precisely, is voltage: the electrical pressure pushing that current along. And the machine that changes voltage, at every single step in that chain, is the transformer.

This is worth sitting with for a moment, because it explains why the transformer is arguably the single most important invention in the history of electric power, more consequential in a narrow but absolute sense than the light bulb, the generator, or even the power plant itself. Without a way to change voltage efficiently, alternating current has no advantage over direct current, and the entire architecture of the modern grid – large power plants serving vast territories through high-voltage transmission – becomes physically impossible. Thomas Edison built his direct-current empire in Manhattan in 1882 without transformers, and it worked, for exactly the reason you would expect a system with no voltage-changing machine to work: badly, and only across short distances. Edison's Pearl Street station could serve customers within about a mile of the generating plant before voltage drop made the system useless. To power a city the size of New York on that model would have required a power plant roughly every square mile, an absurdity that made large-scale electrification simply theoretical.

The transformer solved this problem, and in solving it, decided one of the most consequential technological rivalries of the nineteenth century – the so-called War of the Currents – within about a decade of alternating current's arrival in America. This book's first section tells that story in detail: how a British experimenter tinkering with iron rings in 1831 discovered a physical principle that nobody thought to commercially exploit for over fifty years, how a trio of engineers in Budapest built the first practical transformer in 1885, and how their invention crossed the Atlantic just in time to give George Westinghouse and Nikola Tesla the decisive advantage over Edison.

But this book is not only a history. Its second half is an argument about the present and near future of American energy policy, built on an uncomfortable fact: the same machine that made the twentieth-century grid possible has become, in the 2020s, its single greatest bottleneck. Large power transformers that utilities once ordered and received within six to twelve months now carry lead times of three to five years. The reasons are a tangle of supply chain fragility, a domestic manufacturing base that shrank for decades, a sudden and simultaneous surge in demand from electric vehicles, reshored manufacturing, extreme-weather-driven grid hardening, and – most dramatically – the electricity appetite of artificial intelligence datacenters, which *The Reliability Frontier* described from the utility and policy side without dwelling on the industrial supply chain underneath it. This book dwells on exactly that supply chain: the electrical steel, the copper, the insulating paper and oil, and the small number of skilled winders who can build these machines by hand, because much of the process still cannot be automated.

By the time you finish this book, a gray box in a gravel lot will look different to you. It will look like the machine that let Niagara Falls light Buffalo in 1896, the machine that finally reached the American farmhouse in the 1930s and 1940s, and the machine now standing, quite literally, between a wave of AI datacenters and the electricity they were promised. The story of how it got there begins not with electricity at all, but with a curious Englishman, a coil of wire, and an iron ring.

How This Book Is Organized

Readers should feel free to approach this book in whatever order best suits their interests, though it is written to build cumulatively, with later chapters assuming familiarity with concepts introduced earlier. Part One tells the transformer's invention story, from Faraday's 1831 discovery through the War of the Currents that decided alternating current's commercial triumph by the mid-1890s. Part Two follows the transformer as it built out the American electrical grid, from Niagara Falls through rural electrification and into the mature, largely invisible infrastructure of the mid-twentieth century. Part Three pauses the historical narrative to explain, in plain language, exactly what a transformer is, how the largest units are engineered and manufactured, how they are transported, and what happens when they fail. Part Four turns to the present crisis: an aging fleet, a surge in demand from electric vehicles, reshored manufacturing, and above all artificial intelligence datacenters, and the multi-year manufacturing bottleneck that has resulted, along with the industry's and government's response to it, and a forward-looking assessment of emerging transformer technology.

Readers primarily interested in the contemporary supply-chain crisis and its connection to *The Reliability Frontier* may be tempted to skip directly to Part Four. I would gently discourage this. The engineering explanations in Part Three, and even the historical patterns traced in Parts One and Two, provide essential context for understanding why the current crisis takes the specific shape it does – why, for instance, the bottleneck cannot simply be resolved by manufacturers working overtime, or why the same skilled-labor constraint that slowed transformer production during the Great Depression, discussed in Chapter 8, echoes directly in the workforce challenges manufacturers report today. This book is short enough to read in its entirety in a few sittings, and I believe it rewards that fuller reading.

PART I – THE INVENTION

Chapter 1: Faraday's Ring

In the autumn of 1831, Michael Faraday – a bookbinder's apprentice turned self-taught experimentalist who had risen to run the laboratory at London's Royal Institution – wound two separate coils of insulated copper wire around opposite sides of a plain iron ring, about six inches across. One coil he connected to a battery. The other he connected to a galvanometer, a simple instrument for detecting electric current, with no battery or power source of its own attached. By any intuition available in 1831, nothing should have happened in the second coil. It was not connected, in any direct electrical sense, to the first. There was no wire running between them, only the iron ring they shared.

Faraday closed the switch on the battery circuit and watched the galvanometer needle in the second coil jump, then settle back to zero even though the battery circuit remained closed and current continued flowing steadily through the first coil. He opened the switch, and the needle jumped again, this time in the opposite direction, before again settling to rest. Faraday had just witnessed, and correctly interpreted, electromagnetic induction: a changing magnetic field in one coil could induce a current in a second, entirely separate coil, provided the two were linked by a common magnetic path – in this case, the iron ring that concentrated and carried the magnetic field between them. Critically, the induced current appeared only while the magnetic field was changing. A steady, unchanging current in the first coil produced no effect whatsoever in the second. It was the instant of switching on, and the instant of switching off, that mattered.

This experiment, now sometimes called Faraday's ring or his induction ring, is the direct ancestor of every transformer built since, including the one humming in a substation near your home right now. Faraday himself, characteristically, treated it as one discovery among dozens he made in that extraordinarily productive decade, alongside his work on electrolysis, the discovery of benzene, and his investigations into the relationship between electricity and magnetism generally. He had no obvious commercial application in mind for the induction ring, and indeed no application existed yet, because in 1831 there was no widespread use of electricity to transform in the first place. Faraday was answering a scientific question – does a changing magnetic field produce an electric current? – not solving an engineering problem. The engineering problem would not exist for another half century.

It is worth pausing on why this matters as a pattern, because it recurs constantly in the history of technology and will recur again later in this book: fundamental scientific discoveries and their eventual industrial applications are frequently separated by decades, sometimes generations, and the gap is not usually a failure of imagination on the discoverer's part. Faraday could not have anticipated a nationwide alternating-current grid because nothing resembling one existed to anticipate. What he provided, instead, was the raw physical principle – electromagnetic induction – that later engineers, confronting a real and urgent problem, would recognize as the solution sitting in plain sight in a fifty-year-old physics paper.

Understanding Induction Without the Mathematics

It is possible to understand why Faraday's ring worked, and why it matters, without wading into differential equations. Think of a magnetic field the way you might think of water pressure in a pipe. An electric current flowing through a coil of wire creates a magnetic field around that wire, and if the wire is coiled, the fields from each loop reinforce one another, creating a concentrated magnetic field through the center of the coil, similar to how stacking pressure gauges along a pipe reinforces a pressure reading. An iron ring, unlike air, concentrates and channels a magnetic field extremely efficiently, the way a pipe channels water far more effectively than an open trench. So when Faraday wound his first coil around one side of the iron ring, the magnetic field it produced when current flowed did not stay confined to that side; the iron ring carried it around to the second coil, wound on the opposite side.

Here is the crucial part: a steady magnetic field, no matter how strong, does not push electrons through a nearby wire. It is only a *changing* magnetic field that does that. This is why Faraday's galvanometer needle jumped only at the moments of switching the battery circuit on and off – those were the only moments the magnetic field was actually changing, rising rapidly from zero or falling rapidly to zero. Once the current in the first coil settled into a steady state, the magnetic field it produced stopped changing, and the induced current in the second coil vanished, even though the field itself remained just as strong as before.

This single fact – that only a changing magnetic field induces current – explains something that puzzles many people encountering transformers for the first time: why transformers work with alternating current but are essentially useless with direct current. A battery, or any direct-current source, produces a steady, unchanging current once a circuit stabilizes, and a steady current produces a steady magnetic field, which induces nothing in a second coil. Alternating current, by contrast, is constantly changing – rising to a peak, falling back through zero, reversing direction, rising to a peak in the opposite direction, sixty times a second in the modern American grid. Every one of those sixty cycles per second is, in effect, Faraday's switch being flipped on and off, on and off, continuously and automatically. A transformer built around alternating current does not need a human hand on a switch; the current itself does the switching, endlessly, and the second coil receives an endless, and usable, stream of induced current in response.

This is the physical fact upon which the entire modern electrical grid rests, and Faraday demonstrated it, in its complete essential form, in 1831 – fifty-one years before Thomas Edison opened his Pearl Street generating station, and fifty-four years before anyone built a device recognizable as a practical transformer. The gap between the discovery and its application is the story of the rest of Part One of this book.

Why the World Waited

It is tempting, with the benefit of hindsight, to ask why nobody built a transformer in, say, 1840, once Faraday's principle was published and widely known among physicists. The honest answer is that there was no problem yet for a transformer to solve. The electrical industry of the mid-nineteenth century revolved around telegraphy, electroplating, and small-scale scientific demonstration, none of which required moving large quantities of power over long distances at varying voltages. Telegraph signals carried information, not energy in any meaningful quantity; a telegraph wire could be thin and its current tiny, because the wire only needed to convey a discernible pulse, not power a factory. There was, in short, no electrical *grid* to speak of, no concept of centrally generating power and distributing it to thousands of separate customers, and therefore no reason to solve the specific problem the transformer would eventually solve: getting power efficiently from a distant generating point to many dispersed customers.

That changed only with the rise of electric lighting as a commercial product in the late 1870s and early 1880s. Arc lighting systems, which produced a harsh, brilliant light useful for streets and large public spaces, began to spread through American and European cities in this period, and inventors and entrepreneurs began seriously grappling, for the first time, with the practical challenge of generating electricity at one location and delivering it usefully to many others. It was in this specific commercial context – not in a physics laboratory, but in the messy, competitive world of street lighting contracts and utility franchises – that engineers began rediscovering Faraday's decades-old principle and asking whether it might solve a very current problem: how much of the power generated at a plant was being lost, wastefully, as heat in the wires between the plant and the lamp, and was there a way to reduce that loss.

The answer to that question, worked out independently and nearly simultaneously by several engineers across Europe and North America between roughly 1880 and 1886, turned on a piece of physics almost as elegant as Faraday's induction principle itself, and it is the subject of the next chapter: the relationship between voltage, current, and the wasted heat produced whenever electricity flows through a real wire with real resistance. Understanding that relationship is what finally gave Faraday's fifty-year-old discovery a job to do, and it is worth explaining carefully, because it is the single piece of physics that makes sense of everything else in this book – every giant transformer at a power plant, every gray can on a utility pole, and every headline about utilities struggling in the 2020s to buy the equipment they need.

Faraday's Broader Legacy for This Story

It is worth appreciating just how unusual Faraday was among the scientific figures of his era, because his particular combination of experimental instinct and disinterest in mathematical formalism shaped the induction ring experiment in ways that mattered for everything that followed. Faraday had no formal mathematical training beyond basic arithmetic; he thought in terms of physical "lines of force" threading through space, a visual and intuitive way of picturing electric and magnetic fields that later mathematicians, most notably James Clerk Maxwell, would translate into the rigorous differential equations that now bear Maxwell's name and that underpin the modern theory of electromagnetism. Faraday's ring experiment was not the product of theoretical prediction confirmed by careful measurement; it was the product of patient, methodical experimental tinkering by a man who wanted to know, empirically, whether a magnetic effect could be coaxed into producing an electrical one, having already shown the reverse – that electric currents produce magnetic effects – was true.

This experimental, almost craftsman-like approach to discovery is worth remembering throughout this book, because the transformer's entire subsequent development, from the Hungarian engineers in Chapter 3 through the modern manufacturing challenges in Part Four, has continued to combine rigorous theoretical understanding with a great deal of hands-on, empirical engineering craft – winding coils correctly, selecting the right insulation, testing prototypes to failure to understand their real-world limits. The transformer, unlike some technologies that emerged directly and quickly from pure theory, has always been as much a craft as a science, a characteristic that helps explain, later in this book, why the skilled labor shortage discussed in Chapter 18 is such a genuine and difficult constraint on modern manufacturing capacity: building a transformer well has never been purely a matter of applying a formula, and it still is not.

From Ring to Toy to Tool

In the decades immediately following 1831, Faraday's induction ring found its way into physics lecture halls and demonstration kits across Europe and America, often presented to students as a curious illustration of electromagnetic principles rather than as the seed of a future industry. Induction coils – larger, more powerful descendants of Faraday's original ring, capable of producing high-voltage sparks from a low-voltage battery source – became popular items in traveling scientific demonstrations and, eventually, in early medical and scientific instruments through the middle of the nineteenth century, including early X-ray equipment decades later. These devices exploited the same basic induction principle Faraday discovered, but for entirely different purposes than large-scale power transmission, since, as discussed in Chapter 2, no one yet had a commercial reason to move large quantities of electrical power across long distances.

This intervening period, in which induction technology existed and was reasonably well understood but had not yet found its transformative commercial application, offers a useful reminder for readers encountering emerging technologies in any era: a scientific principle's ultimate significance is very often invisible even to the scientists who discover it, and its eventual application, when it comes, frequently arrives from an entirely different direction – in this case, from the urgent, practical, profit-driven world of electric lighting entrepreneurship in the 1880s, a world Faraday, who died in 1867, never lived to see.

A Note on How to Read the Physics in This Book

Because this book is written for general readers rather than engineers, it is worth being explicit, here at the outset, about the approach it takes to technical explanation throughout. Wherever this book introduces a physical principle – electromagnetic induction in this chapter, the voltage-current-power relationship in the next, or the more specialized engineering concepts introduced in Part Three – it aims to explain that principle using ordinary language and everyday analogy rather than mathematical notation, on the theory that a reader who genuinely understands the underlying logic in plain English has learned something more durable and more useful than a reader who has merely memorized a formula without grasping what it represents physically.

This approach necessarily sacrifices some precision that a formal engineering treatment would include, and readers with technical backgrounds may occasionally find themselves wanting more mathematical rigor than this book provides. That tradeoff is deliberate. The goal of this book is not to train transformer engineers – an outcome that would require years of formal study this book cannot substitute for – but to give an intelligent general reader genuine, accurate insight into why this machine works the way it does, why it was invented when it was, and why its current scarcity matters as much as this book argues it does. Readers seeking deeper technical grounding will find ample references in the further reading section at the end of this book.

Why This Book Starts With Faraday Rather Than With the Grid

Some readers may wonder why a book ultimately concerned with a very contemporary industrial crisis begins nearly two centuries in the past, with a physicist who had no conception of an electrical grid at all. The answer reflects a conviction this book holds throughout: understanding why something matters today requires understanding, concretely, how it came to exist in the first place, and the transformer's origin story is not merely decorative historical color. Every constraint discussed in this book's final section – why transformers cannot be manufactured overnight, why the skilled labor required to build them takes years to train, why the raw materials involved are so specialized – traces directly back to the same fundamental physical principle Faraday discovered in 1831 and the specific engineering solutions later inventors developed to exploit it reliably at scale. Readers who invest the patience to understand that origin story fully will find the contemporary chapters considerably more legible, and considerably more consequential, than they would if this book had simply begun with a description of today's supply chain crisis without first explaining the machine at its center.

Chapter 2: The Problem of Voltage

Imagine two utilities in 1882, each trying to deliver the same amount of power – say, enough to light a thousand incandescent bulbs – from a generating plant to customers a mile away. The first utility sends that power at a low voltage, perhaps 100 volts, the kind Thomas Edison's direct-current systems favored because it was the voltage his light bulbs and generators were designed to use directly, with no intervening machine to change it. The second utility, hypothetically, sends the same amount of power at ten times the voltage, 1,000 volts. Which utility loses more energy, as heat, in the copper wire connecting the plant to the customers?

The answer – dramatically, counterintuitively to most newcomers, and absolutely central to everything in this book – is that the low-voltage utility loses one hundred times more energy to wire heating than the high-voltage utility, for delivering the identical amount of power over the identical distance through identical wire. This is not a minor efficiency difference. It is the difference between a viable business and a hopeless one, and it is the single fact that made Edison's direct-current empire technologically obsolete within roughly a decade of its founding, however brilliant Edison's other innovations were.

Power, Voltage, and Current

To understand why, we need three simple relationships, none of which requires more than arithmetic to grasp. First: electrical power – the actual useful work being delivered, measured in watts – equals voltage multiplied by current. A circuit delivering 100,000 watts of power can do so with high voltage and low current, or low voltage and high current, or any combination in between, so long as the product of the two equals 100,000. This is the crucial degree of freedom that makes transformers valuable: for any given amount of power you need to deliver, you can choose to deliver it at high voltage and low current, or low voltage and high current, and the choice has enormous practical consequences even though the delivered power is identical either way.

Second: the heat wasted in a wire carrying current is proportional not to the current itself, but to the *square* of the current. Double the current flowing through a wire, and you do not double the wasted heat – you quadruple it. Triple the current, and wasted heat increases ninefold. This squared relationship, familiar to physicists as a consequence of Ohm's law and the physics of electrical resistance, is the crux of the entire matter. It means that even modest increases in current produce dramatic increases in wasted energy, while modest decreases in current produce dramatic energy savings.

Third, and following directly from the first two: because power equals voltage times current, delivering a fixed amount of power at ten times the voltage requires only one-tenth the current. And because wasted heat scales with the square of current, one-tenth the current produces only one-hundredth the wasted heat. This is precisely the hundredfold difference in the hypothetical comparison above, and it is not a coincidence or an approximation – it is an exact mathematical consequence of these three relationships working together.

Why Edison's System Could Not Scale

Thomas Edison's Pearl Street station, which began supplying electricity to a square-mile district of lower Manhattan in September 1882, operated at approximately 110 volts direct current – chosen, reasonably enough, because it was a safe and practical voltage for the incandescent light bulbs Edison had spent years perfecting, and because Edison's generators, dynamos of his own design, produced direct current naturally. The system worked, and it worked well, for customers within roughly a mile of the generating station. Beyond that distance, voltage drop and wire losses made the system commercially and technically impractical: the copper required to deliver enough current at 110 volts to serve a customer several miles away, while keeping losses to an acceptable level, would have been prohibitively thick, prohibitively expensive, and in some cases physically impossible to install underground or overhead through existing city infrastructure.

Edison's proposed solution to this limitation was, in essence, to build more power plants – smaller, local direct-current stations scattered every mile or so throughout a city, each serving its own small district, the way Pearl Street served lower Manhattan. This was not an unreasonable engineering response to the constraint direct current imposed; given that DC voltage could not practically be changed after generation, building many small plants close to their customers was the only way to make the arithmetic work. But it was an expensive, capital-intensive, and fundamentally limiting model. It meant that electrifying a large, dispersed metropolitan area, let alone a small town, a rural county, or an entire nation, would require an almost unimaginable number of separate generating stations, each needing its own fuel supply, its own staff, its own boilers, and its own physical footprint, typically within easy reach of the customers it served.

There was, at the time Edison was building his system, no alternative on offer, because no one had yet built a device that could take electricity generated at a convenient voltage – say, whatever voltage a particular generator design happened to produce most efficiently – and change it, cheaply and reliably, to whatever voltage was best suited for transmission over distance, and then change it again, at the far end, to whatever voltage was safest and most convenient for the customer's use. That device, of course, is the transformer, and its absence in the early 1880s was not a matter of missing physics – Faraday had demonstrated the underlying principle fifty years earlier – but a matter of nobody having yet built a practical, reliable, commercially viable machine that exploited it.

Alternating Current's Enabling Property

The reason alternating current, rather than direct current, became the foundation of large-scale electrical grids is now visible in full: alternating current is precisely the kind of continuously changing current that Faraday's induction principle requires to transfer energy between two separate coils sharing an iron core. Direct current, once stabilized, is unchanging, and unchanging currents induce nothing in a neighboring coil, no matter how tightly wound or how well-designed the iron core connecting them. This meant that direct-current systems, as they existed in the early 1880s, had no practical way to change voltage after generation. Whatever voltage a DC generator produced was the voltage that had to be used all the way to the customer, with all the wire losses and distance limitations that entailed.

Alternating-current systems, by contrast, could in principle use a transformer to generate power at whatever voltage suited the generator best, step that voltage up dramatically for the journey across town or across the countryside – minimizing current, and therefore minimizing wasted heat by that all-important squared relationship – and then step the voltage back down, safely, at the point of delivery to a level appropriate for lighting a home or running a factory motor. This was, and remains, the single decisive advantage of alternating current over direct current for large-scale power delivery, and it explains why every national electrical grid in the world today, without exception, is built on alternating current, transformed repeatedly at every stage between generation and use.

It is worth being precise about what "in principle" meant in the early 1880s, because the advantage of alternating current was theoretical until someone actually built a transformer good enough to exploit it commercially. Simply knowing that Faraday's induction principle existed did not automatically produce a device efficient enough, reliable enough, and cheap enough to manufacture at scale for a real electrical utility to depend on. Early attempts at practical transformers, in the first half of the 1880s, were often inefficient, prone to overheating, difficult to manufacture consistently, and in some cases barely more effective than the direct-current alternative they were meant to replace. The theoretical advantage of alternating current, in other words, needed to be matched by an equally serious advance in practical engineering before it could translate into a real commercial victory over Edison's system.

That practical advance arrived within a few years, on the other side of the Atlantic, in a workshop belonging to the Ganz Works electrical manufacturing company in Budapest, where three engineers – Ottó Bláthy, Miksa Déri, and Károly Zipernowsky – built and patented, in 1885, the first transformer design recognizably similar to the machines still manufactured today. Their story, and the specific engineering insight that separated their design from earlier, less successful attempts, is the subject of the next chapter.

A Concrete Numerical Illustration

Because the squared relationship between current and wasted heat is so central to everything that follows in this book, it is worth walking through a single concrete example rather than leaving the claim purely abstract. Suppose a utility needs to deliver 100,000 watts of power to a town five miles from its generating plant, using a copper wire with a fixed resistance of one ohm for that five-mile run. At 1,000 volts, delivering 100,000 watts requires 100 amps of current (since power equals voltage times current, and 100,000 divided by 1,000 is 100). The heat wasted in the wire, calculated as the square of the current multiplied by the wire's resistance, works out to 100 squared, or 10,000, multiplied by one ohm – 10,000 watts wasted as heat, a full ten percent of the power the utility is trying to deliver.

Now suppose the same utility instead transmits that power at 10,000 volts, ten times higher. Delivering the same 100,000 watts now requires only 10 amps of current. The wasted heat becomes 10 squared, or 100, multiplied by one ohm – just 100 watts wasted, one one-hundredth of the loss at the lower voltage, exactly as the mathematics predicts. The utility has delivered the identical quantity of useful power to the identical customers over the identical wire, and reduced its wasted energy from ten percent of total output to one-tenth of one percent, simply by transforming voltage upward for transmission and back downward for delivery. This is not a theoretical curiosity; it is the exact arithmetic that determined, in very concrete dollar terms, whether an 1880s electric utility could turn a profit serving customers beyond the most immediate vicinity of its generating plant.

Why This Math Escaped No One for Long

Once utility engineers and entrepreneurs in the early 1880s began grappling seriously with the economics of electric lighting at any meaningful scale, this relationship between voltage, current, and wasted heat was not a difficult insight to arrive at; the underlying physics, articulated by Georg Ohm decades earlier and refined by subsequent researchers into the power-loss relationship described above, was well established and readily available to any engineer with basic training. The difficult part, as later chapters make clear, was not understanding that higher voltage meant lower losses; it was building a machine that could safely and efficiently change voltage from whatever level a generator produced most efficiently to whatever level was best for transmission, and then change it back down again to a safe level for actual customer use – precisely the problem Faraday's decades-old induction principle was waiting, unrecognized, to solve.

The Illusion of a Simple Fix

It might seem, from the outside, that once utilities understood the voltage-loss relationship described in this chapter, direct-current systems could have simply been redesigned to operate at higher voltage, avoiding the entire alternating-current detour altogether. In practice, this option was far less available than it might appear. Direct-current generators and, critically, the incandescent lamps and early motors that constituted the actual electrical load on Edison-era systems, were designed and manufactured to operate safely and efficiently at the roughly 100 to 110-volt level Edison had selected, a level chosen in part because it represented a reasonable compromise between lamp filament durability, given the materials available in the early 1880s, and worker and customer safety. Redesigning an entire generation and appliance ecosystem to operate at a much higher voltage throughout, simply to reduce transmission losses, would have introduced serious new safety hazards at the point of actual customer use, where high voltage is considerably more dangerous in an ordinary home or business setting than it is along a well-insulated, professionally maintained transmission line running through open countryside or over city rooftops.

This is precisely the problem the transformer solves so elegantly: it allows a system to operate at whatever voltage is safest and most practical at each specific stage – low voltage at the point of generation and appliance use, high voltage only along the transmission path in between, where professional maintenance and physical separation from the public can manage the added risk responsibly. Direct current, lacking any practical way to change voltage between these stages, was permanently locked into a single voltage compromise that could never simultaneously satisfy the competing demands of efficient long-distance transmission and safe customer-facing use. This structural limitation, more than any single design decision Edison made, is what ultimately doomed direct current's prospects as the foundation for a national electrical grid, regardless of how many individual engineering refinements Edison's teams made to their systems throughout the 1880s and early 1890s.

Chapter 3: The Hungarian Three

In the workshop of the Ganz Works electrical manufacturing company in Budapest, in the spring of 1885, three engineers in their twenties and thirties – Ottó Bláthy, Miksa Déri, and Károly Zipernowsky – were wrestling with a problem that had defeated or badly compromised nearly every transformer design attempted up to that point: how to build a device that could handle multiple, independent electrical loads without the voltage at each load fluctuating wildly depending on how many other loads happened to be switched on nearby.

This was not a trivial technicality. It was the difference between a laboratory curiosity and a commercially viable product. The earliest experimental alternating-current transformer systems, including work by the French team of Lucien Gaulard and John Dixon Gibbs demonstrated in London in 1883 and 1884, connected multiple customer loads in *series* along the transformer's secondary circuit – the same electrical arrangement, in essence, as old-fashioned Christmas tree lights, where every bulb shares the same current path and switching one off affects all the others. In a series-connected system, turning off one customer's lamp changed the current flowing through the entire circuit, which in turn changed the voltage delivered to every other customer on that circuit. A system built this way could light a laboratory demonstration convincingly, with a fixed and known number of lamps switched on throughout the test, but it could never serve real customers who turned their lights on and off independently and unpredictably throughout the day and night. Gaulard and Gibbs's system, technically impressive as an early proof that alternating-current transformation was possible at all, was commercially unworkable for exactly this reason.

The Parallel Connection Breakthrough

Bláthy, Déri, and Zipernowsky's crucial insight – patented in Hungary in 1885 and refined over the following two years – was to connect the secondary windings of their transformers to customer loads in *parallel* rather than in series, an arrangement in which each customer's circuit was independent of every other customer's circuit, all drawing from a shared, essentially constant voltage source, the way modern household outlets each draw independently from the same electrical panel without one appliance's operation affecting another's voltage. This might sound like a modest rearrangement of wires, but it was the single design decision that made the transformer a genuinely practical commercial device rather than a laboratory demonstration piece, because it meant a utility could connect any number of customers to a transformer, in any combination of on and off, without one customer's behavior degrading service to another.

Achieving stable parallel operation required more than a change in wiring topology, however. It required a transformer design with high enough internal efficiency and low enough internal electrical losses that the transformer itself did not become the bottleneck, and it required a magnetic core design capable of handling a rapidly and continuously alternating magnetic field without excessive heating or energy loss. The trio's second major innovation addressed exactly this: rather than the simple iron ring Faraday had used, or the straight iron bar cores used in some earlier experimental transformers, Bláthy, Déri, and Zipernowsky developed a closed, ring-shaped core built up from thin sheets of iron, a design that dramatically reduced the wasted energy caused by circulating "eddy currents" induced within the iron itself – unwanted currents that arise, by the same Faraday induction principle used deliberately in the windings, unavoidably within any solid conducting core exposed to a changing magnetic field, converting useful electrical energy into wasted heat within the core material itself.

The laminated core – thin sheets of iron, each electrically insulated from its neighbors by a thin coating of varnish or oxide, stacked together to form the transformer's magnetic path – remains, in modified and improved form, the standard construction technique for transformer cores to this day, over 140 years later. It is a striking example of a fundamental engineering solution arrived at early in a technology's development proving so effective that later generations refine it rather than replace it. Modern transformer cores use grain-oriented electrical steel, a specialized alloy engineered at the metallurgical level to channel magnetic fields more efficiently than the simple iron sheets available to the Ganz Works engineers in 1885, but the basic laminated construction principle – thin insulated sheets rather than a solid block – traces directly back to this period.

The ZBD Transformer and Its Public Debut

The Ganz Works team, whose device became known by the initials of its three inventors as the ZBD transformer, first demonstrated their parallel-connected, closed-core system publicly at the Budapest National General Exhibition in 1885, powering a lighting installation that impressed engineers and journalists alike with its stability regardless of how many lamps happened to be switched on at a given moment. Word of the demonstration spread quickly through the small, tightly networked world of European electrical engineering, and Ganz Works began receiving commercial orders for complete AC lighting systems built around the ZBD transformer design – among the earliest commercial sales, anywhere in the world, of equipment we would recognize today as a functioning power transformer.

Crucially, Bláthy, Déri, and Zipernowsky did not merely build a working prototype; they worked out, with a level of mathematical rigor unusual for the period, the underlying theory of transformer operation, including the relationship between the number of turns of wire in the primary and secondary coils and the resulting ratio between input and output voltage – a relationship so central to transformer operation that it remains the first equation taught in any introductory course on the subject. A transformer with, say, one thousand turns of wire in its primary coil and one hundred turns in its secondary coil will step voltage down by a factor of ten, delivering one-tenth the input voltage to whatever is connected to the secondary side, while correspondingly increasing available current by roughly the same factor (accounting for a small amount of unavoidable internal loss). Reverse the turn counts, with one hundred turns on the primary and one thousand on the secondary, and the same transformer instead steps voltage up by a factor of ten. This turns-ratio relationship, simple as it appears, is the mathematical heart of every transformer built since, from the pole-mounted can outside a suburban home to the largest units at a nuclear power plant switchyard, and it was worked out with essential completeness by the Ganz Works engineers within just a few years of Faraday's principle finally finding an industrial application.

A European Head Start

By the mid-1880s, then, the essential transformer had been invented, patented, and put into limited commercial service in Budapest, roughly contemporaneous with similar – though, as it would turn out, less commercially successful – experiments in Britain and France. It is worth emphasizing how thoroughly this early history is a European story, and specifically a Central European one, because American popular memory of the War of the Currents tends to center almost entirely on Thomas Edison, George Westinghouse, and Nikola Tesla, three men whose contributions were genuinely decisive in bringing alternating current to commercial dominance in the United States but who did not, in fact, invent the transformer that made their systems work. That credit belongs to Bláthy, Déri, and Zipernowsky, and to the somewhat less successful pioneering efforts of Gaulard and Gibbs in London and Paris, whose series-connected system, whatever its commercial shortcomings, helped establish that alternating-current transformation was physically achievable at meaningful scale.

What the ZBD transformer needed, to become more than a European engineering curiosity, was an American champion – someone with the capital, the manufacturing capacity, and the competitive motivation to bring a working alternating-current transformer system across the Atlantic and put it to use in the specific, high-stakes commercial battle then unfolding in the United States over which electrical technology would power the nation's cities. That champion arrived in the person of William Stanley, a young engineer working for George Westinghouse, whose 1886 demonstration in the small Massachusetts town of Great Barrington would mark the true beginning of alternating current's American commercial history – and the subject of the next chapter.

The Fate of the Inventors

The subsequent careers of Bláthy, Déri, and Zipernowsky diverged in ways typical of major collaborative inventions from this period, and worth noting briefly because their story illustrates a pattern that recurs throughout the history of electrical engineering: the individuals responsible for a foundational invention do not always become the household names history remembers. Ottó Bláthy remained at Ganz Works for the rest of his career, continuing to make significant contributions to alternating-current engineering, including important early work on watt-hour meters used to measure electricity consumption for billing purposes – another piece of everyday electrical infrastructure whose obscurity mirrors the transformer's own. Károly Zipernowsky went on to an academic career, eventually becoming a professor at the Budapest University of Technology, training subsequent generations of Hungarian electrical engineers. Miksa Déri continued working in the electrical industry across Europe before his death in 1906.

None of the three achieved the lasting international popular fame of Edison, Westinghouse, or Tesla, despite arguably having done more than any of those three men to invent the specific device – the transformer – without which none of the American War of the Currents developments described in the following two chapters could have occurred at all. This asymmetry between technical contribution and popular historical memory is not unique to the transformer's story, but it is particularly pronounced here, and it is one of the reasons this book exists: to give a comparatively obscure but foundational invention, and its comparatively obscure but essential inventors, the sustained attention their contribution to the modern world deserves.

Ganz Works as an Industrial Pioneer

Beyond the specific technical contribution of its three most famous engineers, Ganz Works itself deserves recognition as one of the pioneering industrial firms of the early alternating-current era, having built and sold complete AC lighting systems – generators, transformers, and associated equipment – to customers across the Austro-Hungarian Empire and beyond within just a few years of the ZBD transformer's 1885 debut. Ganz Works' early commercial success demonstrated something that would matter enormously to Westinghouse's subsequent American ambitions, discussed in the next chapter: that alternating-current lighting systems, transformer included, were not merely laboratory curiosities but genuinely sellable, revenue-generating commercial products capable of competing directly with direct-current alternatives on ordinary commercial terms, a proof of concept that gave Westinghouse's decision to license and import European transformer technology considerably less risk than it might otherwise have carried.

Comparing the ZBD System to Modern Grid Codes

It is genuinely striking, from a modern engineering perspective, how much of the ZBD transformer's basic design logic remains recognizable in the grid interconnection codes and equipment standards utilities enforce today. The parallel-connection principle Bláthy, Déri, and Zipernowsky pioneered – every customer drawing independently from a shared, stable voltage source – remains the fundamental architecture of every modern distribution system, still governed by interconnection standards that trace their conceptual lineage directly back to the 1885 Budapest demonstration described in this chapter. Modern utility engineers rarely have occasion to think about Bláthy, Déri, and Zipernowsky by name when designing a neighborhood distribution circuit, but the architecture they are implementing, at its core, is the same one those three engineers worked out in the Ganz Works laboratory nearly a century and a half ago.

A Note on Simultaneous Invention

The transformer's invention history offers an instructive example of a pattern common throughout the history of technology: important inventions are frequently developed nearly simultaneously, and largely independently, by multiple inventors working in different locations, responding to the same underlying technical and commercial pressures. Alongside the Ganz Works team and the Gaulard-Gibbs efforts in Britain and France discussed in this chapter, other engineers across Europe and North America were experimenting with similar induction-based voltage transformation concepts during the same narrow window of the mid-1880s, reflecting the fact that the underlying commercial problem – efficient long-distance electricity transmission for the emerging electric lighting industry – had become urgent and well-defined enough, by this specific historical moment, that multiple independent engineering teams, drawing on the same fifty-year-old Faraday principle, converged on broadly similar solutions within a remarkably short span of years. This pattern of near-simultaneous, independent invention, common across many major technologies, complicates simple narratives crediting any single inventor with sole responsibility for a breakthrough, and it is part of why this book has taken care to credit Bláthy, Déri, and Zipernowsky's specific, decisive contribution – the parallel-connection, closed-core design that actually proved commercially viable – rather than crediting the more general concept of an alternating-current transformer to any single individual.

Chapter 4: Westinghouse's Machine

George Westinghouse had already made one fortune before he turned his attention to electricity. His railway air brake, patented in 1869 when he was just twenty-two years old, had transformed railroad safety and made him wealthy and well known as an inventor-industrialist in the mold that nineteenth-century America celebrated: a practical engineer who solved a real problem and built a company around the solution. By the mid-1880s, watching Thomas Edison's direct-current lighting systems spread through American cities, Westinghouse became convinced that Edison's approach, however impressive as an initial achievement, contained a fatal structural limitation – the same one described in Chapter 2, the inability to transform voltage after generation – that would eventually cap how far and how cheaply electric lighting could spread.

Westinghouse learned of the European alternating-current transformer work, including the Gaulard-Gibbs system in Britain and, before long, the more commercially promising ZBD design out of Budapest, through the extensive technical press and patent literature that circulated among electrical engineers on both sides of the Atlantic. In 1885, Westinghouse licensed Gaulard and Gibbs's transformer patents and imported several of their transformers, along with a small AC generator, to his manufacturing works in Pittsburgh, intent on determining whether the technology could be made into a viable commercial product for the American market.

William Stanley's Great Barrington Demonstration

The engineer Westinghouse assigned to develop and improve the imported transformer technology was William Stanley Jr., a young inventor who had previously worked on arc lighting systems and joined Westinghouse's company in 1884. Stanley spent much of 1885 and early 1886 redesigning and improving the transformer, correcting inefficiencies in the original Gaulard-Gibbs design and, significantly, adapting the parallel-connection principle that made the Hungarian ZBD system commercially practical, rather than persisting with the series-connection approach that had limited the original European demonstrations.

In March 1886, working from a rented building in the small western Massachusetts town of Great Barrington – chosen partly because Stanley's health required him to work somewhere with cleaner air than industrial Pittsburgh – Stanley built and successfully tested a complete alternating-current lighting system using his improved transformers. The demonstration, though modest by the standards of the vast grids that would follow within a few decades, was carefully engineered to prove the essential commercial claim: a single generator at one location could supply electricity at a convenient generating voltage, a step-up transformer could raise that voltage for transmission along ordinary wire strung through the town, and multiple step-down transformers, installed at various points along the line, could each independently reduce voltage to a safe level for lighting a cluster of nearby buildings, with each transformer's customers switching their lamps on and off without affecting anyone else's service.

Great Barrington's Main Street was lit that spring using this system, making it one of the first, and by most historical accounts the first fully commercial, demonstrations of a complete AC transmission and transformation system in the United States. Westinghouse, watching the results, moved quickly. Within months he had committed his company to a full-scale entry into the alternating-current lighting business, establishing the Westinghouse Electric Company's AC systems as Edison's first serious commercial competitor in the emerging electric lighting industry, and setting the stage for the fierce, sometimes bitter competition between the two approaches that would come to be known, somewhat theatrically, as the War of the Currents.

Why the American Demonstration Mattered

It is worth being precise about what Stanley's Great Barrington system actually proved, because the significance is easy to understate in hindsight, once we already know that alternating current went on to become the near-universal standard. In 1886, this was not yet obvious. Edison's direct-current systems were already installed, functioning, and generating revenue in several American cities. Edison possessed enormous public credibility as America's most celebrated inventor, extensive patents, an established manufacturing and installation infrastructure, and considerable capital. Convincing electric utilities, municipal officials, and the general public that an unfamiliar and, in the mid-1880s, still relatively novel alternative technology deserved serious investment required more than theoretical superiority; it required a working, visible, commercially credible demonstration that skeptics could see with their own eyes.

Great Barrington provided exactly that demonstration, and its timing mattered enormously. Had Stanley's system failed, or had it worked only under carefully controlled laboratory conditions rather than the messy reality of a real small-town Main Street with real, independently operated customer lamps, alternating current's American commercial prospects might have been set back by years, giving Edison's direct-current systems additional time to expand their installed base and entrench their market position before facing serious competition. Instead, the demonstration succeeded, Westinghouse committed decisively to the technology, and within roughly two years Westinghouse Electric was competing aggressively with Edison for lighting contracts in cities across the United States, marketing the specific advantage that alternating current, transformed appropriately at each stage, could deliver electric lighting economically across far greater distances than any direct-current system could match.

The Missing Piece: Motors

One significant gap remained in the Westinghouse alternating-current system as of 1886 and 1887, and Edison's supporters were quick to point to it in the commercial and public relations battle that followed: alternating current, unlike direct current, did not yet have a practical, efficient electric motor design. Direct-current motors were well understood and widely used by the mid-1880s, powering everything from streetcars to factory machinery, and this gave Edison's direct-current systems a genuine commercial advantage for any customer who wanted not just lighting but mechanical power from their electrical connection – a meaningful limitation, since electric lighting alone, however valuable, represented only part of the commercial market utilities hoped to serve.

This gap would be closed remarkably quickly, and by a name closely associated with the alternating-current side of this story ever since: Nikola Tesla. Tesla, a Serbian-American engineer who had briefly worked for Edison before a bitter falling-out over compensation and had struck out on his own, developed and patented a practical polyphase alternating-current induction motor design in 1888, a design that, unlike the AC motor concepts that had been attempted previously, could start and run reliably and efficiently without needing a separate direct-current source for its starting mechanism. Westinghouse, recognizing immediately that Tesla's motor patents solved the single remaining commercial disadvantage of his alternating-current systems, purchased the rights to Tesla's patents later that year, in a licensing arrangement that would eventually become one of the most consequential and, for Westinghouse's company, financially challenging deals in the history of American electrical engineering, as later chapters will discuss.

With Tesla's motor design in hand, alongside Stanley's improved and increasingly efficient transformers, Westinghouse Electric now possessed a complete alternating-current system capable of matching Edison's direct-current offering feature for feature – lighting, motors, and mechanical power – while retaining the singular structural advantage that direct current could never overcome: the ability to transform voltage cheaply and efficiently at every stage between generation and use, enabling transmission across distances Edison's system could never economically reach. The stage was now set for the full-scale commercial and public relations confrontation between the two systems, a conflict that would turn unexpectedly vicious, involve stray dogs, a circus elephant, and the invention of the electric chair, and ultimately be decided as much by engineering economics as by any of the dramatic public spectacles both sides mounted. That confrontation is the subject of the next chapter.

Stanley's Later Career and the Value of a Good Demonstration

William Stanley continued refining transformer design after his Great Barrington demonstration, eventually leaving Westinghouse to found his own electrical manufacturing company, which produced transformers and other electrical equipment into the early twentieth century before being absorbed into General Electric. Stanley's specific technical contributions – improvements to core design and to the practical manufacturability of transformers suited to real commercial deployment rather than laboratory conditions – mattered, but his most consequential act, in the arc of this book's story, may simply have been proving, publicly and visibly, that the imported European transformer concept could work reliably under real American commercial conditions, in an ordinary New England town rather than a controlled demonstration hall.

This point deserves emphasis because it recurs throughout the history of technology adoption generally: a genuinely superior technology does not sell itself automatically merely by being theoretically superior on paper. It requires someone willing to build a working, visible, real-world demonstration that skeptical customers, investors, and competitors can observe directly, and it requires a commercially credible sponsor – in this case, Westinghouse, already a proven industrialist with real capital and manufacturing capability – willing to stake resources and reputation on the demonstration's success. Stanley provided the engineering; Westinghouse provided the credibility and capital to turn that engineering into a serious commercial threat to Edison's established position.

The Westinghouse-Tesla Royalty Renegotiation

The 1888 licensing arrangement between Westinghouse and Tesla for the polyphase induction motor patents, mentioned briefly in this chapter, deserves a bit more context, because it illustrates the financial pressures that alternating-current adoption placed even on its most committed corporate champion. The original royalty arrangement Westinghouse agreed to with Tesla was structured on a per-horsepower basis tied to the eventual scale of AC motor deployment, a reasonable structure at the time it was signed but one that, as alternating current's American market share grew far beyond what either party likely anticipated in 1888, threatened to become an enormous ongoing financial obligation for Westinghouse's company during a period, in the early 1890s, when Westinghouse Electric faced serious financial strain amid a broader national financial panic.

Tesla, according to the widely repeated account of this episode, voluntarily agreed to renegotiate and substantially reduce his royalty claim, reportedly out of personal loyalty to Westinghouse and recognition that the company's continued financial survival was itself essential to alternating current's ultimate commercial triumph over Edison's direct-current systems – a decision that cost Tesla what would likely have been an enormous personal fortune had the original royalty terms remained in force through the following decades of AC motor deployment, but one that arguably helped ensure Westinghouse Electric's survival as the corporate vehicle capable of carrying alternating current, and the transformer technology at its core, to eventual commercial dominance.

Great Barrington Today

It is worth noting, as a small historical footnote, that the site of William Stanley's original 1886 demonstration in Great Barrington, Massachusetts, has been commemorated locally, and the town retains a modest but genuine place in the historical memory of the American electrical industry, even though it never became a major electrical manufacturing center in its own right. This is itself an instructive detail: the location of a foundational technological demonstration does not necessarily become the location of the industry it launches, since the actual manufacturing scale-up that followed Stanley's proof of concept occurred primarily at Westinghouse's existing industrial facilities in Pittsburgh, better suited to large-scale manufacturing than the small Massachusetts town where the crucial proof-of-concept demonstration happened to take place.

Westinghouse's Broader Vision

George Westinghouse's decision to pursue alternating current so aggressively, discussed throughout this chapter, reflected a broader pattern in his career as an inventor-industrialist: a consistent willingness to bet substantial capital on technologies he judged to be fundamentally superior, even when doing so meant challenging entrenched competitors with greater initial market share and public visibility, as his railway air brake business had done decades earlier against established but inferior braking technology. This willingness to make large, technically informed bets against incumbent competitors, informed by genuine engineering judgment rather than mere contrarianism, characterizes much of the most consequential entrepreneurship in the history of American industrial technology, and Westinghouse's alternating-current bet, vindicated decisively by the mid-1890s events described in the next chapter, stands as one of the most successful examples of this pattern in American industrial history – a bet made possible, in the specific and concrete sense this book has emphasized throughout, by the transformer technology Stanley brought successfully across the Atlantic and adapted for genuine commercial deployment.

Chapter 5: The War of the Currents

By 1887, Thomas Edison understood that his direct-current lighting empire faced a genuine commercial threat, and he responded not primarily with engineering improvements to his own system's fundamental distance limitations – there was, after all, no way to give direct current the voltage-transforming ability it structurally lacked – but with an aggressive public relations campaign designed to portray alternating current as unacceptably dangerous. This campaign, often remembered today under the somewhat theatrical label the War of the Currents, is one of the strangest and most ethically uncomfortable episodes in the history of American technology, and it is impossible to understand the transformer's ultimate commercial triumph without understanding it.

Edison's argument, stripped of its theatrics, contained a kernel of genuine technical truth: alternating current, particularly at the higher voltages required for efficient long-distance transmission, was in fact more dangerous in certain failure scenarios than the comparatively low, roughly 110-volt direct current Edison's own systems used. High-voltage alternating current, arcing from a downed line or an improperly insulated piece of equipment, posed a real hazard, and utility linemen in this early, poorly regulated period of electrical installation were indeed dying in accidents involving high-voltage AC lines with troubling frequency, deaths that received considerable and often gruesome newspaper coverage. Edison was not inventing a danger from nothing.

Electrocution as Marketing

What Edison and his associates did with this genuine technical concern, however, went well beyond legitimate safety advocacy. Working with an engineer named Harold Brown, whose relationship with Edison's company was for some time deliberately obscured from the public, Edison's camp staged and publicized a series of demonstrations in which stray dogs, and eventually a circus elephant named Topsy at Coney Island in 1903 (after the main public relations campaign had largely concluded, though the footage remains one of the most infamous artifacts of this era), were killed using alternating current, explicitly to associate AC in the public imagination with instant, gruesome death. Brown lobbied the New York State legislature to adopt alternating current as the method for the state's newly authorized method of capital punishment, a campaign that succeeded in 1888 with the introduction of the electric chair, first used to execute a convicted murderer, William Kemmler, in 1890 – a process that went badly wrong and took repeated attempts to complete, a horrifying spectacle that Edison's camp had nonetheless hoped, cynically, would forever brand alternating current in the public mind with the phrase Edison reportedly encouraged newspapers to adopt: to be "Westinghoused."

This campaign, ugly as it was, ultimately failed to halt alternating current's commercial advance, for reasons that had little to do with public relations and everything to do with the underlying economics described in Chapters 1 and 2. Utilities, city governments, and industrial customers evaluating lighting and power contracts in the late 1880s and early 1890s were, in the end, making capital investment decisions based on which technology could deliver electricity more cheaply and to a larger service territory, not primarily on which technology had won the more effective newspaper campaign. Alternating current's transformer-enabled ability to serve customers many miles from a generating plant, at a fraction of the copper cost required by direct current for the same distance, represented a durable economic advantage that sensationalized demonstrations involving dogs and elephants could not permanently offset, however much they shaped public anxiety in the short term.

The Real Battleground: Distance and Cost

The genuinely decisive engagements in the War of the Currents were fought not in newspaper columns but in specific, real commercial contracts, where the underlying economics played out in front of paying customers rather than a general public audience. A particularly instructive example involves the competitive bidding for the contract to light the 1893 World's Columbian Exposition in Chicago, an enormous and internationally publicized fair that both Edison's General Electric (the company formed from the 1892 merger of Edison's various business interests with the Thomson-Houston Electric Company) and Westinghouse Electric aggressively pursued as a showcase opportunity. Westinghouse won the contract with a substantially lower bid than General Electric's, made possible in significant part by alternating current's ability to serve the fair's sprawling grounds with fewer, more efficient transformers and generating stations rather than the network of closely spaced direct-current plants General Electric's technology would have required. The Exposition, illuminated by some 250,000 incandescent lamps powered by Westinghouse's alternating-current system, became the most visible and effective demonstration yet of AC's practical capability at genuinely large scale, reaching millions of fair visitors in a way no laboratory test or newspaper column ever could.

An even more consequential contract followed close behind: the 1893 award to Westinghouse of the contract to build the generating and initial transmission equipment for the Niagara Falls Power Company's ambitious project to harness the falls' hydroelectric potential and transmit power to the city of Buffalo, New York, some twenty-two miles away – a distance that would have been entirely impractical for any direct-current system of the period to serve economically, and which therefore represented, in the starkest possible terms, the specific commercial application where alternating current's transformer-enabled advantage mattered most. This project, completed in stages through the mid-1890s, is the subject of the next chapter, and it stands as the moment alternating current's theoretical and demonstrated advantages translated into the large-scale, genuinely transformative infrastructure project that set the template for the American electrical grid to come.

Assessing the Verdict

By the mid-1890s, the practical commercial verdict on the War of the Currents was effectively settled, even though direct current never entirely disappeared – it retained real advantages in certain niche applications, including some early streetcar and industrial motor systems, for years afterward, and direct-current transmission technology would eventually return, in a very different high-voltage form, for specific long-distance and undersea transmission applications many decades later, a development touched on again near the end of this book. But for the central commercial question of the era – how should a growing American electric utility industry generate and distribute power to cities, towns, and eventually the whole country – alternating current, and the transformer that made it practical, had won decisively, and would go on to become the essentially universal foundation of electrical grids not just in the United States but around the world.

It is worth stating plainly what this chapter and the three preceding it have collectively established, because it is easy to lose amid the vivid personalities and dramatic incidents of the War of the Currents: the outcome of this contest was determined, in the end, not primarily by the celebrated inventors whose names dominate popular retellings – Edison, Westinghouse, Tesla – but by a quieter technical fact discussed in Chapter 2, and by a specific machine, the transformer, whose essential design was worked out not by any of those three men but by Bláthy, Déri, and Zipernowsky in Budapest and refined for American conditions by William Stanley in Great Barrington. Edison was arguably the greater inventor by any broad measure of his total career achievement, and Tesla's later reputation as an unappreciated genius is not undeserved. But the specific contest chronicled in this chapter was decided by arithmetic: the squared relationship between current and wasted heat, and the transformer's unique ability to exploit that arithmetic by changing voltage cheaply at every stage of a power system. Wherever popular history simplifies this story into a contest of personalities, it obscures the more important and more durable truth – that a comparatively obscure category of machine, more than any individual inventor's genius or salesmanship, decided how electricity would be generated and delivered for the next century and beyond.

That machine's story, however, was only beginning in the 1890s. Part Two of this book follows the transformer as it left the laboratory and the demonstration project behind and became the literal backbone of an expanding, increasingly interconnected American electrical grid – starting at Niagara Falls, where the theoretical case made in this chapter's closing pages was about to be tested at a scale no one had previously attempted.

What the Public Spectacle Obscured

It is worth returning, in closing this chapter, to the electrocution campaign described earlier, because its ultimate commercial failure is instructive in a way that extends well beyond the specific history of electric power. Fear-based public relations campaigns can shape public opinion and even legislative outcomes in the short term – the electric chair's adoption in New York is a direct, documented consequence of Harold Brown's lobbying, undertaken with Edison's quiet support – without ultimately controlling the outcome of a technology competition whose real stakes were commercial and industrial rather than emotional. Utilities, city governments, and industrial customers making multi-year capital investment decisions in the late 1880s and early 1890s were, for the most part, sophisticated buyers evaluating engineering economics, not general members of the public reacting to sensationalized newspaper coverage of dog and elephant electrocutions.

This is not to say public perception was irrelevant; utilities did have to manage public safety concerns about high-voltage lines, and the industry's subsequent development of safety codes, insulation standards, and public education around electrical safety, discussed briefly in later chapters, responded to genuine underlying hazards that Edison's campaign, however cynically deployed, was not entirely wrong to highlight. High-voltage alternating current is genuinely more dangerous than low-voltage direct current in certain failure scenarios, and the industry's subsequent safety engineering took that reality seriously even as it rejected Edison's broader attempt to use that reality to halt AC's commercial advance entirely.

A Note on Historical Memory

Popular retellings of the War of the Currents – including a well-known 2017 feature film that dramatized the rivalry between Edison and Westinghouse – tend to compress this multi-year, multi-front commercial and engineering competition into a more compact and more personality-driven narrative than the historical record actually supports, emphasizing dramatic personal rivalry over the more mundane, but ultimately decisive, economic and engineering realities described in this chapter and the two preceding it. This is a natural consequence of how popular narrative works; a story centered on the arithmetic of resistive power loss and transformer core design does not translate easily into dramatic film or stage narrative the way a story centered on rival industrialists and their competing public demonstrations does. Readers who encounter dramatized versions of this history elsewhere should understand that the underlying commercial verdict, however it is packaged narratively, rested on the specific technical and economic facts this book has laid out in detail across its first five chapters, with the transformer itself, more than any individual person's genius, doing the decisive work.

The Broader Lesson for Technology Competitions

The War of the Currents offers a template worth naming explicitly, because it recurs throughout the subsequent history of technology competition in ways this book's readers will likely recognize from more recent examples: an incumbent technology with an established installed base and considerable public credibility, facing a genuinely superior challenger technology, frequently responds not primarily through direct technical improvement of its own offering – which is often structurally incapable of matching the challenger's core advantage, as direct current could never match alternating current's transformer-enabled distance advantage – but through public relations campaigns, safety concerns real or exaggerated, and appeals to established relationships and reputational trust. These campaigns can meaningfully slow a superior technology's adoption in the short term, as Edison's campaign arguably did to some degree, but they rarely reverse the underlying economic verdict once that verdict is genuinely decisive, as the outcome of this specific historical episode demonstrated clearly by the mid-1890s.

Direct Current's Quiet Return

It is worth noting, as a coda to this chapter, that direct current did not disappear from electrical engineering entirely, and its eventual partial return, decades later, in a very different high-voltage form, deserves brief mention here even though this book examines it more fully in Chapter 20's discussion of emerging transformer and power-electronics technology. High-voltage direct current transmission, made practical only much later through the development of power electronics technology unavailable in the 1890s, has found valuable niche applications for very long-distance transmission and for connecting grids that operate at different frequencies or that are not otherwise synchronized, including some undersea transmission cables and certain long-distance continental transmission corridors. This modern high-voltage direct current technology, ironically, depends on its own specialized category of power-electronics-based conversion equipment that performs, through very different means, much the same fundamental function the transformer performs for alternating current – a reminder that the underlying engineering problem this book's opening chapters describe, moving electrical power efficiently across distance, admits more than one solution, even though alternating current and the transformer decisively won the specific nineteenth-century commercial competition this chapter has examined.

PART II – THE ELECTRICAL AGE

Chapter 6: Niagara Falls

The Niagara Falls Power Company had been trying, in one form or another, to harness the falls' enormous and obvious energy potential since the 1880s, initially through direct mechanical means – water wheels connected by rope or cable drives to nearby mills, an approach with ancient precedent but severely limited reach, since a mechanical drive system could realistically serve customers only within a mile or so of the powerhouse itself. The falls could clearly generate enormous power; the unsolved problem was how to deliver that power usefully to customers who were not conveniently located directly adjacent to the riverbank.

By the time the company's directors seriously evaluated proposals in the early 1890s, the outcome of the previous chapter's War of the Currents had become clear enough that the choice of technology was, in retrospect, barely a contest. A commission that included the eminent British physicist Lord Kelvin reviewed competing proposals, and in 1893 the company awarded the contract for its initial generating and transmission equipment to Westinghouse Electric, whose alternating-current, transformer-based system offered the only realistic prospect of delivering Niagara's power economically to customers beyond the immediate vicinity of the falls.

The Buffalo Transmission Line

The specific engineering challenge facing Westinghouse's team, once the initial generating station began producing power in 1895, was transmitting a meaningful fraction of that power to the city of Buffalo, some twenty-two miles away – a distance that, as discussed in the previous chapter, would have made direct-current transmission prohibitively expensive due to the sheer quantity of copper required to keep voltage-drop losses within acceptable limits. Using step-up transformers at the generating station, engineers raised the voltage of the power produced by Niagara's generators to 11,000 volts for the transmission line to Buffalo, a level that seems almost quaintly modest by the standards of modern extra-high-voltage transmission lines carrying several hundred thousand volts, but which represented, in 1896, an extraordinary leap beyond anything previously attempted commercially for a transmission distance of this length.

The line to Buffalo began operating in November 1896, delivering power that, upon arrival, passed through step-down transformers reducing voltage to levels appropriate for streetcars, industrial motors, and eventually residential and commercial lighting throughout the city. The project's success was immediate and widely celebrated, not merely as an engineering achievement but as tangible, visible proof of exactly the claim alternating-current advocates had been making throughout the previous decade: that a single generating facility, exploiting a naturally advantageous location – in this case, one of the most powerful waterfalls in North America – could serve customers dozens of miles away, at a cost and efficiency direct current's supporters could not begin to match. Newspapers of the period covered the Niagara-to-Buffalo transmission line with the kind of breathless enthusiasm later generations would reserve for the moon landing, and with good reason: it demonstrated, concretely and at commercial scale, the specific architectural principle that would define electrical grids for the next century and beyond – large, centralized generation, connected by high-voltage transmission enabled by transformers, serving dispersed customer populations far beyond what any single local generating plant could reach on its own.

A Template, Not an Endpoint

It would be a mistake, however, to treat the 1896 Niagara-to-Buffalo line as a finished, mature version of the transmission architecture that would eventually span the continent. The transformers Westinghouse installed at Niagara and along the Buffalo line, while dramatically more capable than anything built even five years earlier, were modest by later standards, both in the voltages they handled and in the sheer physical scale of the equipment. Transmission voltage would climb steadily over the following decades as transformer insulation technology, core design, and manufacturing techniques improved, from the 11,000 volts of the original Buffalo line, to 60,000 volts and beyond within a decade or two, to the 230,000, 345,000, and eventually 765,000-volt extra-high-voltage lines that carry the bulk of long-distance power across the American grid today. Each of these voltage increases required, in turn, transformers engineered to handle progressively larger, progressively more demanding insulation and cooling challenges, an engineering progression that forms the core subject of Part Three of this book.

What Niagara Falls established, in a form durable enough to define the following century, was the basic architectural template: generate power where it is most efficient or most naturally advantageous to generate it, transform that power to a high voltage for efficient transmission over whatever distance separates the generating plant from its customers, and transform it back down, at one or more intermediate stages, to whatever voltage is appropriate for the specific customer or class of customer being served. This three-stage architecture – generation, high-voltage transmission, and lower-voltage distribution, with transformers doing the essential work at every transition – remains, in essence, exactly how the American electrical grid operates today, over 125 years later, even as the scale of every individual component has grown by orders of magnitude.

The Utility Industry Notices

The commercial success of the Niagara Falls project reverberated quickly through the still-young American utility industry, whose executives had been watching the War of the Currents' outcome with close professional attention, for reasons that had nothing to do with loyalty to any particular inventor and everything to do with straightforward business calculation. A utility executive in, say, Ohio or Georgia in the late 1890s cared little about which historical figure had invented what; that executive cared about which technology would let his company serve the largest possible customer base, across the largest possible service territory, at the lowest possible capital and operating cost, while remaining reliable enough to avoid the reputational and regulatory damage that outages could bring.

Niagara Falls answered that question in the most persuasive way available to any skeptical executive: not through argument, but through demonstrated, revenue-generating, twenty-two-mile commercial performance. Utility companies across the country accelerated their adoption of alternating-current systems through the late 1890s and into the 1900s, and transformer manufacturing – by this point pursued not just by Westinghouse but by an expanding roster of American electrical manufacturers, including General Electric, which had by then made its own peace with alternating current despite the company's Edison-era direct-current origins – grew into a substantial and increasingly sophisticated industrial sector in its own right.

This growing demand set in motion a process this book's next several chapters will trace in detail: the gradual consolidation of a fragmented, locally organized electric utility industry into larger regional and eventually interconnected systems, a consolidation that both depended upon, and in turn drove further innovation in, the transformer technology at the center of every stage of that expanding grid. That consolidation – and the transformer's quiet, largely unheralded role at the center of it – is the subject of the next chapter. **Adams Power Plant and the Broader Niagara Story**

The Niagara Falls Power Company's efforts extended well beyond the single Buffalo transmission line described in this chapter's main narrative. The company's Adams Power Station, named for company president Edward Dean Adams and completed in stages through the 1890s, eventually housed some of the largest generators built anywhere in the world up to that time, and the transmission and transformation infrastructure built to carry that power outward from the falls expanded steadily over the following decades, eventually serving customers well beyond Buffalo, including industrial customers who located new manufacturing facilities specifically near Niagara Falls to take advantage of the abundant, comparatively cheap electric power the site's hydroelectric generation made available – an early example of a pattern that would recur throughout the twentieth century and into the twenty-first, in which industrial customers cluster their operations near sources of abundant, low-cost electricity, a dynamic directly relevant to the modern datacenter siting decisions discussed in Part Four of this book.

International Attention

The Niagara Falls project attracted international engineering attention from its earliest planning stages, with European engineers and utility executives visiting the site throughout the 1890s to study its generation and transmission technology, and the project's success influenced subsequent hydroelectric and long-distance transmission projects across Europe and elsewhere in the following decades. This international dimension is worth noting because it illustrates that the transformer's story, while this book focuses primarily on its American development and current American supply-chain crisis, has always been a genuinely global one, with European invention, American commercial application, and subsequent international technology diffusion all playing essential roles at different stages – a pattern that recurs, in a very different and more troubled form, in the current global transformer manufacturing supply chain examined in Part Four.

The Financial Structure Behind the Project

The Niagara Falls Power Company's project also deserves mention as an early example of the large-scale project financing structures that would become increasingly important to American infrastructure development throughout the following century. Financing a project of this ambition and technical uncertainty – there was, after all, no precedent anywhere in the world for transmitting this quantity of power this far using this specific technology – required assembling capital from a combination of wealthy private investors, including members of prominent American financial families of the era, and the credibility that came from the involvement of internationally respected engineering authorities, including Lord Kelvin's advisory commission mentioned earlier in this chapter, whose endorsement helped reassure investors that the project's technical approach rested on sound engineering judgment rather than speculative optimism.

This combination of patient private capital and rigorous independent technical review, willing to commit substantial funds to an infrastructure project whose payoff would only be realized over years, offers an instructive contrast to some of the more speculative, rapidly assembled utility holding company financing structures that would emerge in the following decades, discussed in Chapter 7 and examined in greater detail in *A History of the United States Electrical Grid.* The Niagara Falls project's success, both technically and financially, established a template that more responsible utility financing efforts would continue to draw upon throughout the twentieth century, even as other, less careful ventures during the 1920s holding company boom would depart from it, with the consequences that book describes.

Chapter 7: Building the Grid

Readers of my earlier book, *A History of the United States Electrical Grid*, will recognize the broad outline of what happened to the American utility industry between roughly 1900 and 1935: a wave of consolidation, in which small, city-specific utility companies were acquired, merged, and organized into increasingly large holding companies, some of which grew to control electricity generation and distribution across multiple states, eventually provoking the federal regulatory response embodied in the Public Utility Holding Company Act of 1935. That book told this story primarily as a tale of finance, ambition, and regulation. This chapter tells the same period from underneath, as it were – from the perspective of the transformer engineering that made the consolidation not just financially attractive but technically achievable in the first place.

Why Bigger Companies Needed Bigger Transformers

Consolidation is, at its core, a bet that serving a larger territory from fewer, larger generating plants will be cheaper per customer than serving many small territories from many small plants. This bet only pays off, however, if the transmission technology connecting those large plants to the newly acquired, often geographically dispersed territories can move power efficiently across the necessary distances – which is to say, the bet only pays off if transformer technology keeps advancing in lockstep with utility executives' ambitions to expand their service territories.

It did. The period between 1900 and roughly 1930 saw transmission voltages climb steadily, from the tens of thousands of volts typical of the Niagara Falls era to 110,000 volts and beyond by the 1910s and 1920s, each increase driven by the same underlying economic logic explained in Chapter 2: higher transmission voltage means lower current for a given amount of delivered power, and lower current means dramatically lower resistive losses in the transmission wire. Each increase in standard transmission voltage, however, demanded transformers capable of safely handling that higher voltage without insulation failure, arcing, or excessive internal losses – a demand that fell on an expanding domestic transformer manufacturing industry centered around companies like General Electric, Westinghouse, and a growing number of smaller specialized manufacturers.

Meeting this demand required genuine engineering advances, not merely scaling up existing designs. Higher voltages placed far greater stress on the insulating materials separating a transformer's high-voltage windings from its core, its tank, and from each other, and engineers spent the first decades of the twentieth century developing improved insulating oils, improved paper and cloth insulation for the windings themselves, and improved core designs that could handle greater magnetic flux without overheating. The transformer that could safely handle 110,000 volts in 1915 was not simply a larger version of the transformer that had handled 11,000 volts at Niagara Falls in 1896; it incorporated genuinely new engineering solutions to problems – dielectric breakdown, partial discharge, thermal management – that simply had not arisen at the earlier, lower voltage.

Interconnection Begins

As individual utilities' service territories expanded and as neighboring utility systems' transmission lines occasionally came close enough to one another to make physical connection practical, engineers and executives began to recognize a further advantage available from transformer-enabled high-voltage transmission: the ability to interconnect separate utility systems, allowing them to share generating capacity, cover for one another during equipment outages, and smooth out the natural variation in electricity demand across different customer populations and time zones. This interconnection trend, still in its early stages during the 1910s and 1920s, would eventually mature into the large regional interconnections – the Eastern Interconnection, the Western Interconnection, and Texas's separate ERCOT grid – that define the structure of the American electrical system to this day, a structure *A History of the United States Electrical Grid* examined in detail from a regulatory and market perspective.

What is worth adding here, from the engineering side of the story, is that interconnection between separate utility systems is only physically possible because transformers at the boundary between two systems can match voltage levels, and in some configurations, phase relationships, between systems that might have been designed and built by entirely different companies, using different equipment standards, years or decades apart. The transformer's role at these interconnection points – sometimes literally the physical location where two utility companies' equipment meets – made it, quite concretely, the device that allowed the fragmented, locally organized American electrical industry of 1900 to gradually knit itself into the interconnected regional and eventually near-national grid structure that exists today.

Holding Companies and the Limits of Financial Engineering

It is worth noting, in a book primarily concerned with physical engineering rather than corporate finance, that the 1920s holding company consolidation wave chronicled extensively in *A History of the United States Electrical Grid* was not purely, or even primarily, a story of genuine engineering-driven efficiency gains. Some holding companies pursued acquisitions and corporate restructurings for reasons that had more to do with financial leverage, stock promotion, and the era's generally permissive regulatory environment than with any real technical synergy between the utilities being combined – a pattern that would contribute directly to the regulatory crackdown embodied in the 1935 Public Utility Holding Company Act, and to the broader collapse of several major utility holding companies during the Great Depression.

The transformer's story and the holding company's story are related but distinct, and it is important not to conflate them. The genuine technical case for larger, interconnected utility systems – lower per-unit costs through economies of scale, improved reliability through the ability to share reserve generating capacity across a wider territory, and the ability to site generating plants at genuinely advantageous locations like Niagara Falls rather than merely wherever a small local utility happened to already own land – was real, and depended directly on the transformer engineering described in this chapter. But the specific corporate structures built atop that genuine technical foundation during the 1920s often layered financial complexity and leverage that had little to do with engineering efficiency and everything to do with the incentives facing utility holding company promoters during a permissive, poorly regulated financial era. Readers interested in that side of the story should consult the relevant chapters of *A History of the United States Electrical Grid*; this book's concern remains the machine itself, and the machine, whatever the corporate structure surrounding it, kept getting larger, more capable, and more essential throughout this period.

Toward Rural America

By the late 1920s, the transformer technology described in this chapter had built out an increasingly capable and interconnected grid serving America's cities and industrial centers. Rural America, by contrast, remained almost entirely unelectrified, not for lack of technical capability – the same transformers stepping voltage down for a city customer could, in principle, do the same for a farmhouse – but because the economics of serving widely dispersed rural customers, with far fewer customers per mile of line than a dense city neighborhood, discouraged private utility investment in rural service. Closing that gap would require both a specific transformer innovation, suited to the particular economics of serving a single farmhouse rather than a city block, and a major federal policy intervention. Both are the subject of the next chapter. **Three-Phase Power and Its Transformer Implications**

One important technical development this chapter has not yet mentioned, but which shaped transformer engineering throughout the early consolidation era, was the industry-wide shift toward three-phase alternating current for transmission and heavy industrial power, rather than the single-phase systems used in many early demonstrations, including Stanley's original Great Barrington installation. Three-phase power, which delivers electricity through three separate alternating currents offset from one another in timing, offers meaningful efficiency and equipment-size advantages for transmitting large quantities of power and for powering the industrial motors that were becoming an increasingly important electricity customer base as American manufacturing grew through this period.

Adapting transformer design to three-phase systems required its own engineering advances, including three-phase transformer core designs that could handle all three current phases within a single, more compact and more cost-effective unit than three entirely separate single-phase transformers would require, though large three-single-phase-unit banks remained, and remain to this day, a common configuration for certain very large or very high-voltage applications, in part because a bank of three separate single-phase units allows a utility to keep a single spare unit on hand to replace whichever phase might fail, rather than needing an entire spare three-phase unit – a practical consideration directly related to the spare-equipment strategies discussed in Chapter 13.

Regional Variation in Grid Development

It is worth noting that grid consolidation and transformer-enabled interconnection did not proceed at a uniform pace across the country during this period. Densely populated industrial regions in the Northeast and Midwest generally saw more rapid utility consolidation and higher transmission voltages earlier than more sparsely populated regions of the South and West, a regional variation that would persist, in modified form, through the rural electrification era discussed in the next chapter and that continues to shape the somewhat different grid architectures – including the distinctively separate ERCOT grid covering most of Texas – that *A History of the United States Electrical Grid* examines in detail from a regulatory perspective.

Engineering Societies and the Diffusion of Best Practice

The period of grid consolidation and expanding interconnection discussed in this chapter also saw the maturation of professional engineering societies, including what would become the American Institute of Electrical Engineers, founded in 1884 and eventually merging with a related organization to form the modern Institute of Electrical and Electronics Engineers. These professional bodies played an essential, if easily overlooked, role in the transformer's development during this period, providing a forum through which engineers working for competing utilities and manufacturers could nonetheless share technical knowledge, debate emerging design practices, and gradually converge on shared standards for voltage classes, testing procedures, and safety practices – the same standardization process discussed in Chapter 8, examined here from the institutional rather than purely technical angle.

This professional knowledge-sharing infrastructure mattered enormously for an industry whose individual firms were, in a purely commercial sense, often fierce competitors for the same utility contracts and licensing arrangements. The willingness of engineers from competing companies to publish technical papers, present findings at professional society meetings, and gradually converge on shared industry standards reflected a recognition, common across many emerging engineering disciplines of this period, that basic safety and interoperability standards benefited the entire industry, and ultimately the public the industry served, in ways that pure commercial competition alone would not reliably produce. This same institutional pattern – competing firms cooperating through professional societies on baseline standards while competing vigorously on price, service, and specific technical execution – continues to characterize the transformer manufacturing industry today, and it is a meaningful part of why transformers built by different manufacturers, in different countries, can nonetheless be relied upon to meet consistent, well-understood performance and safety benchmarks.

Chapter 8: The Transformer Gets Bigger

If Chapter 7 traced how transformers enabled the geographic expansion of the American electrical grid, this chapter examines a parallel and equally important trend: the transformers themselves, particularly those used at power plants and major substations, grew dramatically larger, more powerful, and more sophisticated between roughly 1900 and the outbreak of the Second World War, in ways that laid essential groundwork for the far larger units that would follow in the postwar decades.

Oil-Filled Cooling

One of the most consequential engineering developments of this period was the maturation of oil-immersed transformer design, in which a transformer's core and windings are submerged in a specially formulated mineral oil, contained within a sealed steel tank. This might seem, at first glance, like an odd or even risky design choice – submerging electrical equipment in a flammable liquid – but the engineering logic is sound and remains the standard approach for large transformers to this day. Transformer oil serves two simultaneous purposes: it acts as an excellent electrical insulator, allowing designers to pack windings more closely together and handle higher voltages within a given physical size than would be possible using air insulation alone, and it acts as an effective cooling medium, absorbing heat generated within the transformer's core and windings and carrying that heat, through natural or forced circulation, to the tank's outer walls or to external radiators, where it dissipates into the surrounding air.

Early transformers, including many of the units installed in the 1890s and early 1900s, often relied on simpler air-cooling or more primitive oil-cooling arrangements that limited how much power a given transformer could handle before internal temperatures rose to damaging levels. Through the 1900s, 1910s, and 1920s, manufacturers steadily refined oil formulations, developed better internal oil-circulation designs, and added external cooling equipment – radiator banks, and eventually fans and oil pumps to force circulation for the largest units – that allowed transformers to handle progressively greater electrical loads within a manageable physical size and without unacceptable degradation of the oil or the paper insulation it protected. This oil-and-paper insulation system, refined over these decades, proved so effective and so durable that many transformers built using 1920s and 1930s-era design principles remained in reliable service for fifty years or more, a durability that has significant and troubling implications for the aging-infrastructure discussion in Part Four of this book.

Rising Voltages, Rising Stakes

As transmission voltages climbed through this period – from roughly 110,000 volts becoming commonplace in the 1910s to 220,000-volt lines appearing by the 1920s and 1930s – the engineering challenge facing transformer designers grew correspondingly more severe. At higher voltages, the insulating materials separating a transformer's windings from each other, from the core, and from the grounded steel tank face dramatically greater electrical stress, and even small manufacturing defects or design miscalculations that might be harmless at lower voltages can produce catastrophic insulation failure at higher ones. Transformer manufacturers responded with increasingly sophisticated insulation systems, more rigorous testing protocols – including high-voltage withstand tests conducted on every unit before it left the factory – and more conservative design margins for their highest-voltage products.

This period also saw the standardization of many design and testing practices across the American transformer manufacturing industry, coordinated in significant part through the work of engineering professional societies and, eventually, formal standards organizations, ensuring that transformers built by different manufacturers, to be installed in different utility systems, would nonetheless share compatible voltage ratings, testing standards, and basic safety margins. This standardization mattered enormously for an industry increasingly characterized by interconnection between separate utility systems, since a transformer manufactured to inconsistent or incompatible standards could become a dangerous weak link at exactly the boundary points where different utilities' systems met.

The Physical Scale Problem

As transformers grew more powerful, they also grew physically larger and heavier – a trend that might seem like a simple, almost trivial consequence of handling more electrical power, but which in fact introduced an entirely new category of engineering and logistical challenge that would only intensify in later decades. A transformer capable of handling the electrical demands of a major 1930s-era power plant might weigh dozens of tons, a scale that already exceeded what could be transported by ordinary road vehicles of the period and required specialized rail cars, designed and built specifically for the unusually heavy, unusually shaped cargo that large transformers represent.

This logistical dimension of transformer manufacturing – getting an enormous, delicate, extraordinarily expensive piece of equipment safely from a factory to its installation site, often hundreds or thousands of miles away – became, during this period, a specialized engineering discipline in its own right, employing custom-designed rail cars called Schnabel cars, specifically engineered to distribute a transformer's immense weight across many axles and to allow the transformer's own body to serve as part of the load-bearing structure during transport. This logistics story would grow only more dramatic in the following decades, as transformers serving the largest power plants and transmission substations grew to weigh several hundred tons, a scale of engineering challenge examined in detail in Part Three of this book.

The Interwar Foundation

By the time the United States entered the Second World War in 1941, the American transformer manufacturing industry had developed most of the fundamental design principles – oil-immersed cooling, laminated silicon-steel cores, paper-and-oil insulation systems, standardized testing protocols, and specialized heavy-transport logistics – that would remain the basic template for large power transformer construction for the rest of the twentieth century and, in modified and improved form, into the twenty-first. The war itself accelerated demand for transformers and electrical equipment generally, as wartime industrial production placed unprecedented demands on the American electrical grid, and the postwar decades that followed would see transformer technology scale to genuinely enormous proportions, driven by an economic boom, a rapidly growing population, and an American lifestyle becoming steadily more dependent on electricity for everything from air conditioning to television.

Before turning to that postwar expansion, however, this book's next chapter examines a specific and, in social terms, arguably the most consequential application of transformer technology in American history: the small, unglamorous pole-mounted or pad-mounted distribution transformer that finally brought reliable electricity to the American farmhouse, closing a gap in rural electrification that had persisted for decades after cities had already come to take reliable electric service for granted.

Bushings, Tap Changers, and Voltage Regulation

Two further engineering developments from this period deserve mention because they remain standard features of large power transformers today. The first is the porcelain and, later, polymer bushing, discussed briefly in Chapter 11, which allows a high-voltage connection to pass safely through a transformer's grounded tank wall; bushing design improved substantially through the 1910s, 1920s, and 1930s as engineers better understood the electrical stress distribution around these components and developed more reliable insulating materials to withstand it. The second is the load tap changer, a mechanical device, developed and refined during this period, that allows a transformer's turns ratio to be adjusted incrementally while the unit remains energized and in service, providing a limited but valuable ability to fine-tune output voltage in response to changing load conditions on the connected system, without needing to take the transformer out of service entirely to make manual adjustments.

The load tap changer deserves particular appreciation because it addresses a real operational challenge: the turns-ratio relationship described in Chapter 3 fixes a transformer's voltage transformation at a specific ratio determined by its winding design, but real electrical systems experience constantly varying load conditions that can cause voltage to sag or rise somewhat from its nominal target level. A tap changer allows grid operators to compensate for these variations by selecting among several slightly different turns-ratio configurations built into the same physical transformer, an elegant engineering solution that remains standard equipment on the vast majority of large power transformers built since this period, and one more example of the kind of incremental but consequential engineering refinement that transformed the basic Faraday-principle device into the sophisticated, reliable machine this book's Part Three examines in full detail.

A Manufacturing Base Takes Shape

By the interwar period, the American transformer manufacturing industry had grown well beyond the two dominant early players, General Electric and Westinghouse, to include a number of additional manufacturers serving different market segments and regions, laying the foundation for the broader, more geographically distributed domestic manufacturing base that would characterize the mid-twentieth century boom discussed in Chapter 10, and whose subsequent contraction, discussed at length in Part Four, lies at the root of today's supply-chain crisis.

The Great Depression's Effect on Transformer Manufacturing

The interwar transformer manufacturing progress described in this chapter did not proceed on an uninterrupted upward trajectory; the Great Depression, beginning in 1929, dramatically curtailed utility capital spending across the industry, as the financial difficulties facing many utility holding companies, discussed in *A History of the United States Electrical Grid*'s treatment of this period, forced widespread deferral of new equipment orders and, in some cases, cancellation of previously planned transmission and generation expansion projects. Transformer manufacturers, like much of American heavy industry, faced sharply reduced order volumes throughout the early 1930s, and the New Deal-era public power programs discussed in that companion volume, including the Tennessee Valley Authority and the Rural Electrification Administration, played an important role not only in expanding electrical access, as discussed in Chapter 9, but also in sustaining transformer manufacturing capacity and expertise through a period when private utility capital spending had contracted sharply.

This historical episode offers a useful, if imperfect, parallel to the current moment discussed in Part Four: it demonstrates that transformer manufacturing capacity, once established, does not simply persist indefinitely regardless of demand conditions; sustained periods of reduced orders can erode manufacturing capacity and specialized workforce expertise in ways that later prove costly to rebuild once demand recovers, a dynamic that operated in the opposite direction – capacity contraction during a demand trough – compared to the capacity contraction during a long period of flat but adequate demand that Chapter 15 identifies as contributing to today's crisis, but which illustrates the same underlying principle: transformer manufacturing capacity is a use-it-or-lose-it resource that requires sustained, adequate demand signals to maintain over time.

Chapter 9: Rural Electrification

In 1935, as detailed in *A History of the United States Electrical Grid*, only about ten percent of American farms had access to electric service, even as the nation's cities enjoyed increasingly reliable, increasingly ubiquitous electrification. That earlier book examined this gap primarily as a story of market failure and government policy response: private utilities, evaluating the cost of extending service to widely dispersed rural customers against the comparatively modest revenue each rural customer represented, generally declined to build the necessary lines, and it took the creation of the Rural Electrification Administration in 1935, along with federally backed low-interest loans to farmer-owned electric cooperatives, to close the gap over the following two decades.

This chapter tells the same story from the perspective of a specific, almost absurdly modest piece of equipment: the small, single-phase, pole-mounted distribution transformer that, multiplied across hundreds of thousands of individual farm connections, actually delivered the electricity that rural electrification promised. It is worth appreciating just how different the economics and engineering of this "last mile" transformer were from the giant units built for power plants and major substations discussed in the previous two chapters, because the difference illustrates a broader and enduring truth about the transformer as a category of technology: it scales, in both directions, across an enormous range, from machines weighing several ounces to machines weighing several hundred tons, and the small end of that range has arguably touched more individual American lives than the large end ever will.

The Economics of the Last Mile

A rural electric cooperative building service to farms scattered across a sparsely populated county faced a fundamentally different economic calculation than an urban utility serving a dense city neighborhood. In a city, a single distribution transformer, mounted on a pole or in a small vault, might serve a dozen or more closely spaced homes, spreading the transformer's cost across many customers. In a rural setting, the same transformer might serve only one or two farms, sometimes located a considerable distance from the nearest neighbor, meaning the fixed cost of the transformer and the line serving it had to be justified by revenue from far fewer customers.

This economic reality shaped both the technical design and the financing structure of rural electrification. On the technical side, engineers working for rural cooperatives and for manufacturers serving the rural electrification market developed smaller, cheaper single-phase distribution transformers specifically optimized for the light loads typical of an early rural customer – perhaps a handful of light bulbs, a radio, and eventually a refrigerator or a water pump, a far smaller electrical demand than a typical city household, let alone the industrial and commercial customers that anchored many urban utility systems' revenue. These smaller rural transformers, often rated at just a few kilovolt-amperes, a small fraction of the capacity of a typical urban distribution transformer, allowed cooperatives to serve individual or small clusters of farms without the capital expense of oversized equipment designed for loads the rural customer had no realistic prospect of using, at least initially.

On the financing side, the federal loan programs administered through the Rural Electrification Administration provided the low-cost, long-term capital that made building these lines and transformers economically viable for farmer-owned cooperatives in a way it had not been for profit-seeking private utilities evaluating the same rural territories against alternative, more immediately profitable investments elsewhere in their service areas.

A Transformer at Every Farmhouse

The physical result of this combined technical and financial effort, playing out across the late 1930s, the 1940s, and into the early 1950s, was the installation of literally millions of small pole-mounted transformers across rural America, each one typically serving just one or a handful of farms, stepping the several-thousand-volt distribution voltage carried along the cooperative's rural lines down to the 120 and 240 volts a farmhouse's wiring and appliances required. By 1953, roughly ninety percent of American farms had electric service, a transformation completed within less than two decades of the Rural Electrification Administration's creation, and one of the most dramatic and rapid infrastructure buildouts in American history.

It is worth pausing on what this transformation actually meant for the households it reached, because the significance goes well beyond convenience.

Electric lighting eliminated the fire risk and constant maintenance burden of kerosene lamps. Electric pumps brought running water into farmhouses that had previously relied on hand-drawn wells, transforming daily domestic labor, particularly for women, in ways historians of rural life have documented extensively. Electric refrigeration transformed food storage and, over time, farm economics, allowing perishable products to be preserved and transported in ways previously impossible. Radio, and eventually television, connected isolated rural communities to national culture and information in ways that had simply not been possible without electric power. None of this – not one bulb, not one refrigerator, not one radio – could have reached a rural American household without the specific, modest, and largely unheralded engineering achievement of a small transformer mounted on a pole beside a farmhouse, quietly doing the same fundamental job as the largest generating-station transformer at Niagara Falls: taking electricity at one voltage and delivering it, safely and usefully, at another.

A Universal Pattern

The rural electrification story illustrates a pattern that recurs throughout the transformer's history and that this book returns to repeatedly: the same underlying physics and the same basic engineering principle – electromagnetic induction, exploited through a carefully designed magnetic core and set of windings – scales across an extraordinary range of applications, from a small ceramic-cased device weighing a few pounds and serving a single farmhouse, to a four-hundred-ton behemoth serving an entire regional transmission grid. What differs across this range is not the underlying physics, which remains essentially the same equation Bláthy, Déri, and Zipernowsky worked out in Budapest in 1885, but the engineering details of insulation, cooling, core design, and manufacturing precision required to build a transformer safely and efficiently at each specific scale.

This scalability is, in a sense, the transformer's quietest but most remarkable achievement: it is simultaneously one of the largest, most expensive, most logistically complex pieces of equipment in the entire electrical grid, and one of the smallest, cheapest, most universally distributed. Both extremes rest on the same physical principle, and both extremes, as this book's remaining chapters will make clear, face distinct but related versions of the supply-chain and manufacturing-capacity crisis discussed in Part Four.

By the middle of the twentieth century, then, the transformer had completed its transformation, so to speak, from a scientific curiosity into an utterly essential, and largely invisible, piece of everyday American infrastructure – as present in the electrified farmhouse as in the largest power plant. That invisibility, and the engineering maturity that produced it, is the subject of the next and final chapter of Part Two.

The Cooperative Business Model

The farmer-owned electric cooperative structure that emerged to carry out much of the rural electrification effort deserves attention in its own right, because it represented a genuinely distinctive institutional innovation, developed specifically to make the challenging economics described in this chapter work. Unlike investor-owned utilities, which needed to generate returns for outside shareholders, member-owned cooperatives could operate on a not-for-profit, cost-of-service basis, charging members only enough to cover the cooperative's actual operating and debt-service costs, an institutional structure well suited to serving a customer base with the thin, geographically dispersed revenue potential rural customers represented in the 1930s and 1940s.

This cooperative structure, still very much alive today – many rural Americans continue to receive electric service from member-owned cooperatives descended directly from Rural Electrification Administration-era organizing efforts – illustrates a broader theme relevant throughout this book: the transformer itself is a politically and institutionally neutral piece of technology, equally capable of serving a customer whether that customer is served by an investor-owned utility, a municipal utility, or a member-owned cooperative. What determined whether rural customers actually received transformers and the electric service they enabled was not any technical limitation of the equipment itself, but the financing and institutional structures available to fund their installation – a reminder that the history this book tells is never purely a story of engineering, but always also a story of the economic and institutional arrangements that determine where and how engineering gets deployed.

Legacy Equipment Still in Service

A striking number of the small distribution transformers installed during the original rural electrification buildout of the late 1930s through the early 1950s remained in service for remarkably long periods, in some documented cases for well over fifty years, a testament to the durability of the oil-and-paper insulation systems described in Chapter 8, but also a contributing factor to the aging-fleet concerns examined in Part Four of this book. Many rural cooperatives, serving customer bases that grew only slowly if at all through the second half of the twentieth century, had less financial pressure and less operational urgency to replace these aging units than utilities serving faster-growing urban and suburban territories, meaning rural distribution infrastructure in some areas carries a disproportionate share of the oldest transformers still in active service on the American grid today.

Measuring Success Beyond Percentages

The ninety-percent farm electrification figure cited earlier in this chapter, while a genuinely remarkable statistic, understates the full scope of what rural electrification accomplished, because it counts only whether a connection existed, not how thoroughly electricity transformed the economic structure of rural life once that connection arrived. Electric-powered milking machines, feed grinders, and water pumps did not merely make existing farm labor more convenient; they fundamentally reshaped how much labor a single farm family could productively manage, contributing to the broader twentieth-century consolidation of American agriculture into larger, more mechanized operations, a transformation with consequences still shaping American rural life and agricultural economics today.

The transformer's role in this transformation, while indirect compared to the tractors, milking machines, and other electrified equipment that farmers actually purchased and operated, was nonetheless essential and foundational: none of that equipment could function without the reliable electric service the small pole-mounted transformers described in this chapter made possible, delivered at a price rural cooperatives could sustain thanks to the federal financing structures discussed above. It is a useful reminder that infrastructure's significance is often best measured not by the infrastructure itself, but by the entire ecosystem of economic and social activity it makes possible once it is in place – a theme this book's Part Four returns to directly when discussing what might be lost, economically and socially, if the current transformer bottleneck constrains the next wave of American electrification for years longer than necessary.

Chapter 10: The Machine Becomes Invisible

By the 1950s and 1960s, the American electrical grid had reached a level of engineering maturity and reliability that would have seemed almost miraculous to the engineers who built the Niagara Falls transmission line just sixty years earlier. Electricity, once a novel and somewhat mysterious commercial product, had become an assumed background feature of American life, as unremarkable and as taken for granted as running water or paved roads. This chapter examines how that transition happened, and argues that the transformer's near-total disappearance from public awareness during this period is not a sign of the technology's diminishing importance, but precisely the opposite: a sign of how completely it had succeeded at its job.

The Paradox of Excellent Engineering

There is a broader pattern worth naming here, because it applies to the transformer more than almost any other piece of major infrastructure equipment, and because understanding it helps explain why so few Americans today could correctly describe what a transformer does, despite interacting with the consequences of its operation dozens of times a day. Technologies that fail often or visibly draw sustained public attention and scrutiny; technologies that work reliably, silently, and consistently over long periods tend to recede from public consciousness precisely because they give people no reason to think about them. A bridge that collapses becomes a subject of intense public discussion; a bridge that carries traffic safely for eighty years becomes, over time, simply part of the landscape, unremarked upon by the people who cross it daily.

The transformer, by the postwar period, had become exactly this kind of successful, unremarked-upon infrastructure. The engineering advances chronicled in the previous chapters – oil-immersed cooling, laminated silicon-steel cores, increasingly sophisticated insulation systems, and increasingly rigorous manufacturing and testing standards – had produced a device so reliable that the average American encountered transformer failure, if at all, as a rare and localized inconvenience rather than a recurring feature of daily life. A pole-mounted distribution transformer failing and causing a neighborhood outage became, by the 1950s and 1960s, an unusual enough event that it merited a phone call to the utility company and, often, a brief period of mild inconvenience rather than the kind of persistent, expected unreliability that had characterized much less mature electrical systems decades earlier.

Postwar Expansion at Unprecedented Scale

None of this reliability should be mistaken for stagnation. The postwar decades, from roughly 1945 through the early 1970s, saw the American electrical grid expand at a pace and scale that dwarfed even the ambitious growth of the interwar period, driven by a rapidly growing population, a booming manufacturing economy, and an American consumer lifestyle increasingly built around electric appliances – air conditioning, television, and eventually a growing array of household electronics – that dramatically increased per-household electricity consumption compared to the more modest lighting-and-basic-appliance loads typical of the 1930s and 1940s.

Meeting this surging demand required an enormous buildout of new generating capacity, including the first generation of commercial nuclear power plants beginning in the late 1950s and accelerating through the 1960s and 1970s, and a corresponding buildout of transmission capacity to move that power from increasingly large, increasingly centralized generating plants to the customers who needed it. Transmission voltages climbed further during this period, with 345,000-volt lines becoming common by the 1950s and 1960s and 500,000-volt and even 765,000-volt extra-high-voltage lines entering service by the late 1960s and 1970s for the longest and highest-capacity transmission corridors. Each of these voltage increases, as in earlier decades, required transformer manufacturers to solve new insulation, cooling, and manufacturing challenges, pushing the largest power transformers to sizes and capacities that would have seemed almost fantastical to the engineers who built the original Niagara Falls system in the 1890s.

Standardization and Maturity

This period also saw the transformer manufacturing industry, along with the broader electrical equipment industry, achieve a level of standardization and manufacturing maturity that made large-scale grid expansion logistically manageable. Industry standards organizations, including what would become the Institute of Electrical and Electronics Engineers and the American National Standards Institute, along with equipment testing and certification bodies, developed detailed, widely adopted standards governing transformer design, testing, and performance, ensuring that utilities across the country could specify, order, and install transformers from multiple competing manufacturers with confidence that the equipment would perform to consistent, well-understood standards.

This standardization, combined with a mature and geographically distributed domestic manufacturing base – including major facilities operated by General Electric, Westinghouse, and a growing roster of other manufacturers across the country – meant that American utilities during this period could generally order large power transformers and expect delivery within a matter of months, an ordering and delivery timeline that stands in stark and troubling contrast to the multi-year lead times utilities face in the 2020s, a contrast this book's final section will examine in detail.

A Machine Doing Its Job Too Well

It is worth closing Part Two of this book by naming directly the pattern this chapter has been building toward: the transformer's near-total disappearance from American public consciousness by the latter half of the twentieth century was not an accident of poor public relations or insufficient cultural significance. It was the direct, almost inevitable consequence of the transformer having become so reliable, so well-standardized, and so consistently available that an entire domestic manufacturing and utility industry could simply assume its continued availability, the way one assumes the continued availability of clean tap water or paved roads, without much conscious thought about the manufacturing capacity, the raw materials, or the skilled labor required to keep that assumption true.

This assumption held, largely unchallenged, for most of the second half of the twentieth century and the first two decades of the twenty-first. It has begun, in the 2020s, to come apart – not because the underlying engineering has failed, but because the manufacturing capacity that once made abundant, reliably available transformers an unremarkable fact of American infrastructure life has shrunk, aged, and struggled to keep pace with a sudden and largely unanticipated surge in demand. Understanding exactly how and why that capacity shrank, and what the machine itself actually looks like inside, requires first taking the time this book's Part Three devotes to genuinely understanding the transformer as a piece of engineering – its core, its windings, its insulation, and the reasons the largest units among them can weigh as much as four hundred tons and take years to manufacture and deliver. That closer examination begins in the next chapter.

Household Electricity Consumption Transformed

It is worth quantifying, briefly, just how dramatically postwar household electricity consumption grew, because that growth drove much of the transformer and grid capacity expansion this chapter describes. A typical American household's electricity consumption grew several-fold between the late 1940s and the early 1970s, driven by the widespread adoption of electric water heaters, clothes dryers, dishwashers, and, especially in the South and Southwest, air conditioning, which transformed both regional population patterns – enabling much larger year-round populations in hot-climate states like Florida, Texas, and Arizona than had been practical before widespread air conditioning – and the seasonal shape of electricity demand itself, introducing the pronounced summer peak-demand pattern that continues to shape utility planning and, as *The Reliability Frontier* discusses extensively, reliability risk to this day.

Every one of these new household electrical loads ultimately passed through a distribution transformer serving that specific home or neighborhood, and utilities during this period engaged in continuous, systematic distribution system upgrades, replacing smaller, older transformers installed during earlier decades with larger units capable of serving the dramatically increased per-household electricity consumption this postwar lifestyle shift produced – an upgrade cycle that, unlike the current crisis described in Part Four, generally proceeded smoothly, supported by the abundant domestic manufacturing capacity and the standardized equipment specifications this chapter has described.

A High-Water Mark

In retrospect, the several decades following the Second World War may represent something of a high-water mark for the American transformer manufacturing and grid-expansion ecosystem: manufacturing capacity was abundant and geographically distributed, engineering standards were mature and well-established, demand growth, while substantial, proceeded at a pace utilities and manufacturers could reasonably plan for years in advance, and the underlying Faraday-principle technology itself, refined over the preceding sixty years, had reached a level of reliability that made large-scale, confident grid expansion possible without the kind of acute equipment scarcity that defines the current period. Understanding exactly what changed, and why that comfortable equilibrium proved impossible to sustain indefinitely, is the task of this book's concluding section.

The Cold War Backdrop

It is worth noting that the postwar grid expansion described in this chapter unfolded against the backdrop of the Cold War, which shaped American infrastructure planning in ways that extended to transformer and grid engineering specifically. Concerns about the electrical grid's vulnerability to potential attack, including the electromagnetic pulse effects that a nuclear detonation could theoretically produce, prompted early research into grid hardening and equipment resilience that, while never resulting in the kind of comprehensive nationwide hardening program some contemporary studies recommended, nonetheless established an early foundation for the physical and, later, cyber-security thinking discussed in Chapter 14. This Cold War-era attention to infrastructure resilience represents an early instance of a pattern that recurs throughout this book: the transformer's centrality to the entire electrical system means that its vulnerabilities, whatever the source, carry consequences far beyond the equipment itself, a fact grid planners have understood, in one form or another, for the better part of a century.

Setting the Stage for Deregulation

The mature, reliable, largely invisible transformer infrastructure this chapter describes also set the stage, inadvertently, for the deregulation era that would begin in the 1990s and that *A History of the United States Electrical Grid* examines in detail across several chapters. Precisely because the grid had become so reliable and so thoroughly taken for granted by the late twentieth century, policymakers and economists began asking whether the traditional, vertically integrated, cost-of-service regulated utility model – developed in significant part to ensure the kind of careful, well-funded infrastructure investment this chapter has described – remained necessary, or whether market competition could deliver comparable reliability at lower cost. That question, and the consequential answer different states gave it, shaped the financial incentives around transformer fleet replacement discussed in Chapter 15, and represents one of the clearest points of connection between this book's engineering narrative and its companion volume's regulatory and financial narrative.

PART III – THE MACHINE

Chapter 11: What Is Inside a Transformer?

Having traced the transformer's history from Faraday's laboratory ring to the largely invisible, largely trusted infrastructure of the mid-twentieth century, it is time to open one up – figuratively, if not literally, since a real power transformer is a sealed, oil-filled steel tank that no reader should attempt to open without specialized training and equipment – and examine, in plain language, exactly what is inside, and why each component is built the way it is.

The Core

At the heart of every transformer sits the core: a structure built from thin sheets of specially formulated electrical steel, stacked and interleaved to form a closed magnetic loop, typically shaped like a rectangular window frame or a set of overlapping rectangular loops, depending on the specific design. The core's job is to provide an efficient magnetic path linking the primary and secondary windings, concentrating the magnetic field produced by current in the primary winding and channeling it through the secondary winding, exactly as Faraday's iron ring did in 1831, but engineered, in a modern power transformer, to handle vastly more power with vastly less wasted energy.

The steel used in modern transformer cores is not ordinary structural steel; it is grain-oriented electrical steel, an alloy containing a small percentage of silicon, processed through a specialized rolling and annealing process that aligns the steel's internal crystal grain structure in the direction the magnetic field will travel through the core. This grain alignment dramatically reduces the energy wasted as the magnetic field repeatedly reverses direction sixty times per second – a phenomenon called hysteresis loss – and, combined with the thin lamination and insulating coating between each sheet described in Chapter 3, also minimizes the wasted energy from eddy currents circulating within the core material itself. Producing this specialized electrical steel requires a small number of highly specialized steel mills worldwide, capable of the precise rolling, annealing, and coating processes the material demands – a manufacturing bottleneck this book's final section will examine as one of the central constraints on expanding transformer production capacity today.

The Windings

Wrapped around sections of the core are the transformer's windings: coils of copper or, in some designs, aluminum conductor, insulated with layers of specially treated cellulose paper and, in many designs, further insulated and mechanically supported by pressboard structures engineered to withstand both the electrical stresses of normal operation and the substantial mechanical forces that can arise during short-circuit events, when currents can briefly spike to many times their normal operating level and the resulting magnetic forces can attempt to physically deform or displace the windings if they are not robustly braced and supported.

A power transformer typically contains at least two separate windings – a primary winding, connected to the incoming electrical circuit, and a secondary winding, connected to the outgoing circuit – wound around the same core section so that the magnetic field produced by current in the primary winding links effectively with the secondary winding. As explained in Chapter 3, the ratio of turns between the primary and secondary windings determines the transformer's voltage ratio: a transformer designed to step voltage down from 138,000 volts to 12,000 volts, for instance, will have roughly eleven and a half times as many turns of wire in its primary winding as in its secondary winding, a ratio derived directly from the desired voltage transformation.

For very large power transformers handling extra-high voltages, the winding design becomes an intricate exercise in managing electrical stress, because the insulation separating adjacent turns, and separating the winding as a whole from the core, the tank, and other windings, must reliably withstand not just normal operating voltage but the much higher transient voltage spikes that lightning strikes and switching operations on the connected transmission system can occasionally produce. Engineering this insulation system correctly, with sufficient margin to survive decades of both normal operation and occasional extreme transient events, is one of the most demanding aspects of large power transformer design, and it is a major reason large transformers cannot simply be manufactured quickly by unskilled labor using generic materials; the insulation design and the precision winding process require specialized engineering expertise and, in many cases, skilled technicians who have spent years learning the craft of winding these coils correctly by hand.

Oil and Cooling

As described in Chapter 8, the core and windings of most large power transformers are submerged in mineral oil (or, in a growing number of modern installations, a less flammable synthetic ester fluid) contained within a sealed steel tank. This oil serves the dual function of providing high-quality electrical insulation between components and carrying heat away from the core and windings toward the tank's outer surface or external radiator banks, where it dissipates into the surrounding air, sometimes assisted by cooling fans and oil circulation pumps for the largest, highest-capacity units.

The oil itself requires careful engineering and ongoing maintenance attention throughout a transformer's operating life. Over decades of service, transformer oil can gradually degrade, absorb moisture, or accumulate byproducts of insulation aging, and utilities routinely test oil samples from their transformer fleets to monitor for early signs of internal problems – a practice called dissolved gas analysis, which can detect certain characteristic gases produced by overheating or electrical arcing inside a transformer well before the problem becomes severe enough to cause an outright failure. This kind of condition monitoring has become an increasingly important tool for utilities managing an aging transformer fleet, a topic Part Four of this book examines in detail.

Bushings and the Tank

Completing the basic anatomy of a power transformer are the bushings: specialized, heavily insulated fittings that allow the high-voltage electrical connections to pass safely from the transformer's internal windings, through the grounded steel tank wall, to the external transmission or distribution lines connected to the transformer. Bushings must provide reliable electrical insulation across a substantial voltage gradient within a compact physical space, and they represent, in practice, one of the more common failure points in an operating transformer fleet, a fact utilities' maintenance programs specifically account for.

The steel tank itself, beyond simply containing the oil, core, and windings, often incorporates a pressure-relief system, and increasingly, monitoring sensors that track internal oil temperature, oil level, and other operating parameters, feeding data back to utility control centers that can flag developing problems before they escalate to failure. Modern large power transformers, in this sense, are not the purely passive, unmonitored devices of a century ago; they increasingly incorporate real digital sensing and communication capability, a trend this book's final chapter, on the transformer of the future, examines in more detail.

A Machine of Deceptive Simplicity

Understood at this level of detail, the transformer reveals itself as a device of deceptive simplicity: the basic underlying principle, electromagnetic induction between two coils sharing a magnetic core, is exactly what Faraday demonstrated in 1831, and remains entirely unchanged in every transformer built since. What separates a modern extra-high-voltage power transformer from Faraday's simple iron ring is not a different physical principle, but an enormous accumulated body of materials science, insulation engineering, precision manufacturing, and quality assurance practice, refined over nearly a century and a half, that allows the same basic principle to be exploited safely and reliably at voltages and power levels Faraday could scarcely have imagined.

This accumulated engineering complexity is precisely why large power transformers cannot simply be manufactured quickly, cheaply, or by inexperienced firms attempting to enter the market opportunistically in response to a supply shortage – a fact with direct and serious consequences for the transformer bottleneck this book's final section examines. Before turning to that contemporary crisis, however, the next chapter examines the largest and most demanding category of transformer built today: the extra-high-voltage units that anchor the American transmission grid, and the specific engineering challenges that emerge only at the very largest scale.

Core Geometry: Core-Form and Shell-Form Designs

Beyond the basic laminated-core principle described above, transformer engineers distinguish between two broad core geometries, each with distinct advantages for different applications. In a core-form design, the windings surround the core's vertical legs, with the core forming a simple rectangular loop; this design is comparatively simple to manufacture and remains common for many power transformer applications. In a shell-form design, the core instead surrounds the windings on multiple sides, providing better mechanical support and, in some designs, improved short-circuit withstand capability, at the cost of somewhat greater manufacturing complexity. Large power transformer manufacturers typically offer both design families, selecting between them based on the specific voltage class, power rating, and mechanical requirements of a given order, and the choice between core-form and shell-form design represents one of the earliest and most consequential decisions in the custom-engineering process described in Chapter 12.

Why Aluminum Sometimes Replaces Copper

Chapter 18 discusses aluminum winding substitution as a response to copper supply constraints, but it is worth explaining briefly, here in the core technical chapter, why the substitution is not simply a drop-in replacement. Aluminum conducts electricity less efficiently than copper for a given cross-sectional area, meaning an aluminum winding must be physically larger than a copper winding to carry the same current with comparable losses – a tradeoff that is often manageable for lower-voltage, larger-physical-footprint applications like some distribution transformers, but which becomes more difficult to accommodate in the most compact, most demanding extra-high-voltage designs, where minimizing physical size while managing electrical stress is already a delicate balancing act, as Chapter 12 describes. Aluminum also has different thermal expansion and mechanical properties than copper, requiring engineers to re-verify, rather than simply assume, that a design's mechanical bracing and short-circuit withstand characteristics remain adequate when the winding material changes – another example of how even seemingly simple material substitutions ripple through a transformer's entire engineering design in ways that require real engineering time and testing to validate properly.

Dissolved Gas Analysis in Practice

The dissolved gas analysis technique mentioned above deserves a slightly fuller explanation, given how central it has become to modern transformer fleet management. When a transformer's internal insulation or oil is exposed to abnormal heat or electrical arcing, even at a low level well short of outright failure, the affected materials break down chemically in ways that release specific gases – including hydrogen, methane, ethylene, and acetylene, among others – into the surrounding oil, where they remain dissolved at concentrations a laboratory can measure precisely from a small oil sample. Different gases, and different ratios between them, are associated with different underlying problems: certain gas signatures suggest overheating of the core or windings, while others suggest electrical arcing or partial discharge activity within the insulation system. By tracking these gas concentrations over time through periodic oil sampling, utility engineers can often identify a developing internal problem well before it progresses to outright failure, providing a valuable early-warning tool that has become standard practice across the utility industry and that plays an increasingly important role in the predictive-maintenance strategies discussed in Chapter 20.

Nameplate Ratings and What They Actually Mean

Every transformer carries a nameplate, fixed to its tank, listing its key electrical specifications: rated power, typically expressed in kilovolt-amperes or megavolt-amperes, its rated primary and secondary voltages, its impedance, and various other technical parameters relevant to the engineers who will install and operate it. It is worth explaining briefly why transformer capacity is conventionally expressed in volt-amperes rather than simple watts, a detail that occasionally confuses readers encountering transformer specifications for the first time. Volt-amperes measure what electrical engineers call apparent power – the straightforward product of voltage and current – while watts measure real power, the actual useful work being delivered, and the two can differ depending on the electrical characteristics of the connected load, a phenomenon related to what engineers call power factor. Because a transformer's windings and insulation must be sized to handle the actual current flowing through them regardless of the connected load's power factor, transformer capacity is conventionally rated in volt-amperes, a reflection of the physical current-handling limit the equipment's construction actually imposes, rather than in watts, which would understate the equipment's necessary sizing for loads with less favorable power factor characteristics.

Cooling Class Designations

Modern transformer nameplates also typically include a cooling class designation, a coded abbreviation describing the specific cooling method or methods the unit employs, since many large transformers are designed with multiple cooling stages that engage progressively as load increases, allowing a single unit to handle a range of loading conditions efficiently. A designation such as ONAN, for instance, indicates oil-natural-air-natural cooling, in which oil circulates by natural convection and heat dissipates to surrounding air without forced fans or pumps, appropriate for lower loading conditions; many large units supplement this baseline cooling method with forced-air and forced-oil-circulation stages, engaging cooling fans and oil pumps automatically as internal temperature sensors detect rising load, extending the transformer's usable capacity considerably beyond what natural convection cooling alone could safely support. This tiered cooling approach allows utilities to specify transformers that operate efficiently, with minimal fan noise and energy consumption, during typical loading conditions, while still retaining substantial additional capacity margin for the peak-demand periods, discussed throughout *The Reliability Frontier*, when the full cooling system engages to handle maximum safe loading.

Why Standardized Testing Protocols Matter So Much

Returning to the testing regime discussed briefly in this chapter and more fully in Chapter 12, it is worth emphasizing why standardized testing protocols, developed and refined over decades by industry bodies, matter so much to the reliability record this book has described. A transformer purchased from a manufacturer in one country, tested according to internationally recognized standards, can be installed with confidence by a utility on the opposite side of the world, secure in the knowledge that the unit has been verified to perform to a consistent, well-understood set of electrical and thermal benchmarks. This standardization, easy to take for granted, represents decades of accumulated engineering consensus about exactly what tests meaningfully predict long-term reliable performance, and it is a quieter but no less important achievement of the transformer industry than any single design innovation discussed elsewhere in this book.

Synthetic Ester Fluids and the Push for Safer Insulation

Alongside traditional mineral oil, discussed earlier in this chapter, an increasing number of new transformer installations, particularly in locations with heightened fire-safety or environmental sensitivity, use synthetic or natural ester-based insulating fluids instead. These ester fluids offer a substantially higher fire point than mineral oil, meaning they are considerably more resistant to ignition in the event of an internal fault severe enough to breach the tank, an important safety advantage for transformers installed near occupied buildings or environmentally sensitive areas. Natural ester fluids, derived from vegetable oils, additionally offer improved biodegradability in the event of a leak or spill, an increasingly important consideration for transformers installed near waterways or other sensitive environmental areas.

These ester fluids come with tradeoffs, including generally higher cost and, in some formulations, somewhat different cooling performance characteristics compared to traditional mineral oil, requiring careful engineering evaluation of whether a specific installation's fire-safety and environmental benefits justify the added cost and design considerations involved. Utilities have increasingly specified ester-filled transformers for installations in populated areas, near sensitive water resources, or in jurisdictions with fire codes that specifically favor higher-fire-point insulating fluids, a trend that illustrates how transformer design continues to evolve incrementally, even within the well-established oil-immersed cooling paradigm described throughout this chapter, in response to safety, environmental, and regulatory considerations that did not carry the same weight when the basic oil-immersed transformer design matured during the decades discussed in Part Two of this book.

Chapter 12: The Giant Transformers

At the very top of the size and complexity range described in the previous chapter sit a category of equipment the industry generally calls large power transformers – a somewhat bland engineering term for machines that can stand three stories tall, weigh as much as four hundred tons including their oil fill, cost anywhere from two million to well over ten million dollars depending on voltage class and specifications, and take, in the current market environment, three to five years to design, manufacture, test, and deliver. These are the transformers found at major power plants, stepping generator output up to transmission voltage, and at the largest transmission substations, stepping extra-high-voltage transmission power down for regional distribution, or connecting one transmission voltage level to another at points where different parts of the grid meet.

Why Size Itself Becomes an Engineering Problem

It might seem, on first consideration, that building a larger transformer is simply a matter of scaling up the same basic design described in the previous chapter – more core steel, more copper winding, a bigger tank, more oil. In practice, size introduces genuinely new engineering challenges that do not appear, or appear only mildly, at smaller scales, and understanding these challenges is essential to understanding why the largest transformers are so expensive, so specialized, and so difficult to manufacture quickly.

The first such challenge involves heat. As a transformer's physical dimensions increase, its volume – and therefore its capacity to generate internal heat from core losses and winding losses – grows roughly with the cube of its linear dimensions, while its surface area, which determines how effectively it can shed that heat to the surrounding environment, grows only with the square of its linear dimensions. This means that, all else equal, larger transformers face a fundamentally harder cooling problem than smaller ones, relative to their own heat-generating capacity, and require increasingly sophisticated cooling systems – larger and more numerous external radiator banks, forced oil circulation using internal pumps, and banks of cooling fans, sometimes with multiple stages that engage automatically as load and internal temperature increase – to keep internal temperatures within safe operating limits.

The second challenge involves electrical insulation at extra-high voltage. As described in the previous chapter, insulation design becomes progressively more demanding as voltage increases, not in a simple linear fashion but in ways that require careful engineering analysis of electric field distribution throughout the transformer's internal structure, to ensure no single point experiences electrical stress beyond what the insulating materials at that specific location can reliably withstand over decades of service. At the extra-high-voltage levels used by the largest transmission transformers – 230,000, 345,000, 500,000, and 765,000 volts – this insulation design work requires specialized engineering expertise, sophisticated computer modeling, and extensive high-voltage testing that smaller, lower-voltage transformer designs simply do not require to the same degree.

The third challenge, related to the first two but distinct from them, involves the sheer mechanical forces a large transformer's internal structure must withstand, particularly during short-circuit events on the connected power system, when fault currents can briefly reach many times the transformer's normal rated current, producing correspondingly large magnetic forces that attempt to physically deform the windings. Designing a winding and support structure that can survive these forces, repeatedly, over decades of service without gradual mechanical degradation, requires careful structural engineering analysis that becomes progressively more demanding as the transformer's overall size and power rating increase.

Custom Engineering, Not Mass Production

A crucial fact about large power transformers, essential to understanding both their cost and the supply-chain crisis examined later in this book, is that they are generally not mass-produced, off-the-shelf products in the way that, say, a household appliance or an automobile is. Each large power transformer is typically engineered to specific requirements set by the utility purchasing it – the exact voltage transformation required, the specific power rating needed, the particular electrical characteristics of the location where it will be installed, and often specific requirements related to the physical space available at the installation site, seismic considerations, and the particular environmental conditions the unit will face.

This custom-engineering approach means that ordering a large power transformer is closer to commissioning a custom-built piece of industrial machinery than to purchasing an item from a catalog. A utility places an order specifying detailed technical requirements, the manufacturer's engineering team develops a specific design meeting those requirements, that design typically undergoes review and approval by the utility's own engineers, and only then does actual manufacturing begin – a process that, even before the supply-chain pressures discussed later in this book, could take a year or more from initial order to delivery for the largest, most demanding specifications.

Testing Before Delivery

Before a large power transformer leaves the factory, it undergoes an extensive battery of tests designed to verify that it meets its specified electrical performance and can safely withstand the voltage stresses, including transient overvoltage events, it will encounter throughout its operating life. These tests typically include measurements of the transformer's electrical losses under various load conditions, verification of its voltage ratio and impedance characteristics, and high-voltage withstand tests that intentionally subject the unit to voltage levels well above its normal operating voltage, simulating the kind of transient stress a lightning strike or switching event might produce, to confirm the insulation system can reliably survive such events throughout the transformer's operating life.

This testing regime, standardized through industry bodies including the Institute of Electrical and Electronics Engineers and international counterparts, exists because a large power transformer that fails after installation is an extraordinarily costly and disruptive event – not merely because of the equipment's own replacement cost, but because of the outage and reliability consequences of losing a major piece of transmission or generation infrastructure, sometimes for months, while a replacement is engineered, manufactured, and delivered. The testing regime's rigor is, in a real sense, an insurance policy against exactly this kind of costly failure, purchased at the cost of additional time and expense during manufacturing.

A Small Number of Qualified Manufacturers

Because of the combined demands described in this chapter – specialized engineering expertise, access to specialized materials including grain-oriented electrical steel, sophisticated manufacturing facilities capable of handling components weighing hundreds of tons, and rigorous testing capability – the number of manufacturers worldwide capable of producing the largest, highest-voltage power transformers is relatively small, and the number of factories within the United States capable of doing so is smaller still, a fact of direct and central importance to the supply-chain crisis this book's final section examines in detail. Understanding why so few factories can build these machines, and why building new factory capacity takes years rather than months, requires first understanding one more dimension of the challenge these giant machines present: simply getting them, once built, from the factory to their installation site – a logistical undertaking dramatic enough to merit its own chapter, and the subject of the one that follows.

How Utilities Specify a Custom Transformer

The custom-engineering process described in this chapter typically begins with a utility's own planning engineers determining the specific electrical requirements a new or replacement transformer must meet: the exact voltage transformation needed, the power rating required to serve current and reasonably foreseeable future load at the installation site, the specific impedance characteristics needed to coordinate properly with the rest of the connected electrical system's protective relay settings, and any site-specific physical constraints, including available space, seismic design requirements in earthquake-prone regions, and noise limitations for installations near residential areas, since large transformers, particularly those with substantial cooling fan systems, produce an audible hum that can become a genuine neighborhood concern if not properly managed through design and site placement.

This detailed specification document, often running to dozens of pages for the largest and most demanding units, is then sent out for competitive bid to qualified manufacturers, each of which develops its own detailed engineering design meeting the specification before submitting a bid that includes both price and, critically for the supply-chain crisis discussed in Part Four, a proposed delivery timeline. Utilities evaluating competing bids must weigh not just price but delivery timeline and manufacturer reliability, a calculation that has grown considerably more consequential as delivery timelines have stretched from months to years, as described in Chapter 17.

Impedance and Its Practical Importance

One technical characteristic worth explaining, because it recurs in utility engineering discussions throughout this book, is a transformer's impedance – a measure of how much the transformer itself resists the flow of current, distinct from the voltage-transformation function described in earlier chapters. A transformer's impedance value affects how much fault current will flow through the transformer during a short-circuit event elsewhere on the connected system, which in turn affects how the utility's protective relay equipment, described in Chapter 14, must be configured to detect and respond to such events safely. Because impedance must coordinate correctly with the broader protective system architecture at a specific installation site, a replacement transformer cannot simply be swapped in from a different location or a different utility's spare inventory without careful engineering review to confirm its impedance characteristics remain compatible with the local system's protection scheme – a constraint that limits how freely utilities can share or repurpose spare transformer capacity, even within the pooled spare-equipment programs discussed in Chapter 13.

Seismic and Environmental Engineering

Transformers installed in earthquake-prone regions, including much of the American West Coast, require additional structural engineering to ensure the unit's internal components, tank, and bushings can withstand significant ground motion without catastrophic failure, a design consideration that adds further engineering complexity and, often, cost to units destined for these regions. Similarly, transformers installed in coastal or otherwise corrosive environments require specialized tank coatings and material selections to resist long-term corrosion, and transformers installed in extreme cold climates require oil formulations and heating systems engineered to maintain adequate oil fluidity and insulation performance even at very low ambient temperatures – each representing another dimension of the genuinely custom, site-specific engineering process that distinguishes large power transformer manufacturing from more standardized industrial production.

The Economics of a Single Unit

It is worth pausing to appreciate the sheer capital intensity a single large power transformer represents, both for the utility purchasing it and for the manufacturer building it. A large extra-high-voltage unit, fully engineered, manufactured, tested, and delivered, can represent a capital expenditure running from several million to, for the very largest and most demanding specifications, well over ten million dollars – a substantial sum for a single piece of equipment, though one that must be understood in the context of the many decades of service life such a unit is expected to provide, and the enormous quantity of electricity it will carry over that service life. Utilities generally evaluate these purchases using long-term financial planning tools that spread the capital cost across the equipment's expected decades of useful service, but the immediate cash outlay required, particularly for a utility facing several such purchases simultaneously amid the demand surge described in Part Four, represents a genuine financing challenge layered atop the lead-time and availability challenges this book's final chapters describe.

Redundancy Design and the "N-1" Planning Standard

Utility transmission planning conventionally follows what engineers call an "N-1" reliability standard, meaning the system must be able to continue operating safely and reliably even if any single major piece of equipment – including any single large power transformer – unexpectedly fails or must be taken out of service for maintenance. Meeting this standard typically requires either genuine redundant transformer capacity at critical substations, so that the loss of one unit does not leave the substation unable to serve its full connected load, or sufficiently robust interconnection with neighboring parts of the grid that power can be rerouted around the affected substation without unacceptable service degradation, an approach that depends on the broader interconnected grid architecture discussed in Chapter 7.

Achieving N-1 redundancy at every critical substation across an entire utility's service territory is expensive, since it generally means installing more total transformer capacity than any single moment's peak demand actually requires, in order to maintain that safety margin against unexpected equipment loss. The transformer bottleneck described throughout this book's final section has made maintaining this redundancy standard considerably more difficult and expensive than it once was, since utilities facing multi-year lead times for replacement equipment have correspondingly less flexibility to take a redundant unit temporarily out of service for planned maintenance, or to quickly restore full N-1 redundancy after an unplanned failure has consumed a substation's built-in reserve margin – another way in which the manufacturing bottleneck examined throughout Part Four translates into concrete, measurable reliability risk for the grid as a whole, beyond its more visible effects on new customer interconnection timelines.

Extra-High-Voltage Direct Current Converter Transformers

A specialized category of large power transformer worth mentioning, given the brief discussion of high-voltage direct current transmission in Chapter 5, involves the converter transformers used at the terminals of high-voltage direct current transmission links, which must handle the specific electrical characteristics of the power-electronic conversion equipment on the direct-current side of the link, including harmonic distortion and other electrical stresses not present in ordinary alternating-current-to-alternating-current transformer applications. These converter transformers represent some of the most specialized and demanding engineering applications in the entire large power transformer industry, generally manufactured by an even smaller subset of qualified manufacturers than ordinary extra-high-voltage transformers, and often carrying correspondingly longer lead times and higher costs, a niche but important illustration of how the general manufacturing capacity constraints discussed throughout Part Four can be even more acute for the most specialized transformer applications.

Generator Step-Up Transformers as a Special Case

Another specialized application worth distinguishing is the generator step-up transformer, installed directly adjacent to a power plant's generating units to raise generator output voltage to transmission voltage, as first demonstrated at the Niagara Falls project described in Chapter 6. These units must be engineered to coordinate precisely with the specific electrical characteristics of the generator they serve, including its rated voltage, its behavior during fault conditions, and its physical arrangement at the power plant site, and they are typically ordered and installed as part of an integrated power plant construction project rather than as a standalone utility equipment purchase, meaning their lead times and delivery schedules must coordinate closely with the broader multi-year timeline of building an entire new generating facility – a coordination challenge that has grown more complex as the transformer lead times described in Chapter 17 have stretched to durations approaching or exceeding some power plant construction timelines themselves, occasionally making transformer delivery, rather than the generating unit's own construction, the critical-path item determining when an entire new power plant can actually begin operating.

Chapter 13: Moving a 400-Ton Machine

A large power transformer, once manufactured and tested, presents a logistical challenge almost as demanding as the engineering challenge of designing and building it in the first place. A unit weighing three or four hundred tons, standing several stories tall, and containing delicate internal windings and insulation structures that can be damaged by excessive tilting, shock, or vibration, must somehow travel from a factory – frequently located overseas, given the shrunken state of domestic manufacturing capacity discussed in Part Four – to a power plant or substation site that may be hundreds or thousands of miles away, often located at the end of rural roads never designed to carry loads of this size and weight.

Rail: The Schnabel Car Solution

For transformers traveling significant overland distances within the United States, specialized railcars called Schnabel cars remain the preferred solution, a technology first developed in the 1930s and refined considerably since. A Schnabel car is engineered so that the transformer itself becomes part of the load-bearing structure of the railcar: rather than resting the transformer atop a conventional flatcar, a Schnabel car's structure attaches directly to specially designed lifting points on the transformer's own tank, distributing the enormous weight across a longer wheelbase and more axles than a conventional railcar could provide, and lowering the overall center of gravity to keep the load stable through curves and over uneven track.

Even with this specialized equipment, moving a transformer by rail requires extensive advance planning: surveying the entire rail route to confirm bridges, tunnels, and track curvature can accommodate the load's weight and dimensions, coordinating with railroad operators to schedule the move during periods that minimize disruption to regular rail traffic, and often securing special permits and escort arrangements for the sections of the journey. A single large transformer's rail journey can take days or weeks to complete, moving at greatly reduced speed compared to ordinary freight, with careful monitoring throughout to detect any signs of shifting load or structural stress.

Barge and Ship Transport

For transformers manufactured overseas – an increasingly common scenario given the state of domestic manufacturing capacity – the journey typically begins aboard a specialized heavy-lift cargo ship, capable of loading and securing extraordinarily heavy, awkwardly shaped cargo for an ocean crossing that can take several weeks depending on the transformer's origin and destination port. Upon arrival at a U.S. port, the transformer may continue its journey by barge along navigable inland waterways, where barge transport, unconstrained by the axle-weight limitations that govern rail and road transport, can often handle even greater weights more easily than rail, provided the destination site has practical river or coastal access.

Not every power plant or substation, of course, enjoys convenient access to navigable water or a rail line capable of handling Schnabel car transport, and a significant number of transformer deliveries require at least a final leg of transport by specialized heavy-haul trucks, capable of carrying loads far beyond what an ordinary commercial truck could manage, but still limited by road weight restrictions, bridge capacity, and the physical geometry of highway interchanges and rural roads that were never designed with four-hundred-ton loads in mind.

Route Planning as Its Own Engineering Discipline

Planning the overland portion of a large transformer's journey has become a specialized engineering and logistics discipline in its own right, sometimes requiring surveys of every bridge, overpass, and sharp turn along a proposed route months or even years before the actual transformer is ready to ship, to confirm the infrastructure along that route can safely bear the load and that the vehicle carrying it can physically navigate every turn without striking overhead structures, utility lines, or roadside obstacles. In some documented cases, utilities and their logistics contractors have needed to temporarily remove and later reinstall traffic signals, utility poles, or highway signage along a planned route, or have identified the need for temporary reinforcement of a bridge, months in advance of the actual transformer shipment, specifically to accommodate a single delivery.

This planning burden adds meaningfully to the overall time required to get a newly manufactured transformer from factory to service, on top of the multi-year manufacturing lead time discussed in Part Four, and it represents a cost and complexity that many people outside the utility and heavy-logistics industries simply never have occasion to learn about, since a successfully completed transformer delivery is, almost by design, an unremarkable, quietly executed logistical achievement that draws public attention only in the rare cases when something goes wrong.

Security and Vulnerability in Transit

The specialized, low-speed, carefully choreographed nature of large transformer transport also creates a window of physical vulnerability that utility security planners take seriously, particularly in the years since the 2013 sniper attack on Pacific Gas and Electric's Metcalf transmission substation near San Jose, California – an incident examined in detail in the next chapter – raised industry-wide awareness of the physical security risks facing critical transformer infrastructure, both while installed in the field and, to a lesser degree, during the vulnerable window of transit between factory and installation site.

The N-1 Problem and Spare Transformer Strategy

Because a large power transformer takes so long to manufacture, test, and deliver, utilities have long recognized the strategic value of maintaining spare transformers – either fully assembled spare units held in reserve, or at minimum, critical spare components that can be used to repair or partially rebuild a failed unit more quickly than manufacturing an entirely new one from scratch. Industry-wide spare transformer sharing programs, including a national initiative called the Grid Assurance program and an earlier industry effort known as STEP (Spare Transformer Equipment Program), have attempted to create pooled reserves of critical spare transformers that participating utilities can draw upon in the event of an unexpected failure or attack, reducing the risk that any single utility might face months or years without a critical piece of transmission infrastructure while waiting for a replacement to be manufactured through the normal, multi-year supply chain.

These spare-equipment programs, valuable as they are, address only part of the underlying vulnerability, because the pooled reserve of spare transformers available industry-wide remains small relative to the total fleet of transformers in service, and replenishing a spare that gets deployed to replace a failed unit still requires going through the same multi-year manufacturing process discussed throughout this book's final section. The next chapter examines exactly what happens, and has happened, when this vulnerability is tested by real transformer failures – both the routine mechanical and electrical failures that occur across any large equipment fleet, and the rarer but more dramatic failures caused by extreme weather, cyberattack, or deliberate physical attack – and why these failure scenarios have become an increasingly urgent concern for grid planners across the country.

A Documented Journey

To make the logistics described in this chapter more concrete, consider the kind of documented multi-modal journey that large power transformers destined for interior American utilities regularly undertake: manufactured at an overseas factory, the unit is first transported by specialized internal factory crane and heavy-haul trailer to a nearby port, where it is loaded, using massive dockside cranes rated for hundreds of tons, onto a specialized heavy-lift cargo vessel. After an ocean crossing that can take two to four weeks depending on origin and destination, the transformer is offloaded at a U.S. port – often one specifically equipped with heavy-lift crane capacity, since not every American port can handle cargo of this weight and size – and then transferred either to a barge for further travel along navigable inland waterways, or directly onto a Schnabel railcar or heavy-haul trailer for the overland portion of its journey.

Utilities and their logistics contractors typically begin route-planning work for this overland portion of the journey many months, and in some documented cases years, before the transformer itself is even manufactured, surveying every bridge, tunnel, overpass, and sharp curve along multiple candidate routes to identify the path that best accommodates the load's weight and dimensions while minimizing the need for temporary infrastructure modifications. In some cases, utilities have needed to coordinate with state transportation departments to temporarily raise or relocate overhead utility lines, traffic signals, or highway signage along the planned route, obtaining special transport permits that can take weeks or months to process given the load's exceptional size, before a single mile of the actual overland journey begins.

When Transformers Cannot Reach Their Destination Easily

In a small number of documented cases, utilities have determined that no practical overland route exists to deliver an intact, fully assembled transformer to certain especially remote or geographically constrained installation sites, requiring the alternative of manufacturing and shipping the transformer in modular components that can be reassembled on site – a more expensive and logistically complex approach that adds its own engineering and quality-assurance challenges, since final assembly and testing must then occur at the installation site itself rather than under controlled factory conditions, but which represents the only practical option when a site's road, rail, and waterway access simply cannot accommodate a fully assembled unit of the required size.

The Cost of Getting It Wrong

Transformer transport failures, while rare given the extensive planning this chapter describes, carry catastrophic consequences when they do occur: a transformer that shifts, tips, or is otherwise damaged during transit can suffer internal winding displacement or insulation damage that may not be immediately visible externally but that can cause the unit to fail, sometimes catastrophically, once energized and placed into service – a failure mode that underscores why the extensive factory testing described in Chapter 12 is sometimes supplemented by additional testing after a long or difficult transport journey, before a utility commits to energizing an expensive, long-lead-time unit that has just completed a demanding multi-week, multi-modal transit. **A Global Logistics Network Built for Rare Cargo**

Because large power transformers ship so infrequently relative to ordinary industrial cargo, the specialized heavy-lift shipping and Schnabel-car rail transport infrastructure described in this chapter is maintained by a genuinely small number of specialized logistics firms worldwide, each with deep, hard-won institutional experience in exactly the kind of route surveying, permitting, and load-securing engineering this chapter has described. This concentration mirrors, in its own way, the manufacturing concentration discussed in Part Four: just as only a handful of factories worldwide can build the largest transformers, only a handful of logistics firms can reliably move them once built, and a utility planning a major transformer delivery must typically book this specialized transport capacity well in advance, coordinating carefully with the manufacturer's own production and testing schedule to ensure transport capacity is available when the unit is actually ready to ship.

This dual scarcity – scarce manufacturing capacity and scarce specialized transport capacity, each requiring advance booking measured in months or years – adds yet another layer of lead time to the overall journey from initial order to energized, in-service transformer, beyond the manufacturing lead times discussed at length in Chapter 17. Utilities and manufacturers have, in recent years, begun coordinating this transport planning earlier in the overall project timeline than was historically necessary, recognizing that transport capacity constraints, while generally less severe than manufacturing capacity constraints, can nonetheless add weeks or months of additional delay if not planned for well in advance.

International Comparisons in Transformer Logistics

It is worth noting briefly that the specific transport challenges described in this chapter are considerably less severe in some other countries than in the United States, largely due to differences in inland waterway and rail network density and capacity. Countries with more extensive inland waterway networks navigable by heavy barge traffic, or with rail networks specifically engineered from the outset to handle heavier axle loads than much of the American freight rail network was originally designed for, can sometimes move large transformers with somewhat less elaborate route-specific engineering than the American logistics challenges this chapter has described. This is not to suggest American transformer logistics are uniquely difficult – the specialized Schnabel car and heavy-lift shipping techniques described in this chapter work reliably and have moved large transformers safely to nearly every corner of the country for decades – but simply to note that the specific logistics challenge facing American utilities reflects, in part, broader characteristics of American transportation infrastructure that differ meaningfully from those of some other industrialized nations, a comparison worth keeping in mind when American commentators sometimes describe the United States' transformer delivery timelines as uniquely troubled.

Chapter 14: When Transformers Fail

Every large power transformer, however well engineered, will eventually fail – through gradual insulation aging, an internal fault, an external event like a lightning strike or short circuit that exceeds its design margins, or deliberate attack. Because the transformer occupies such a central position in the grid's architecture, described throughout this book, its failure modes and the grid's resilience against them deserve careful attention, both as a matter of engineering interest and because they illuminate exactly why the current supply-chain bottleneck, discussed in Part Four, represents such a serious reliability concern rather than merely a commercial inconvenience for utilities.

The 2003 Northeast Blackout

The largest blackout in North American history, on August 14, 2003, offers an instructive, if somewhat indirect, illustration of how transformer and transmission equipment failures can cascade into consequences vastly disproportionate to their initial, individual scale. The blackout, which left roughly fifty million people across the northeastern United States and Ontario, Canada, without power, some for as long as four days, originated not from a transformer failure per se, but from a combination of a software bug in an alarm system at a Midwestern utility control center, several transmission lines contacting overgrown trees and tripping offline in rapid succession, and a cascading series of automatic protective equipment operations – including protective relays designed to disconnect transformers and other equipment automatically to prevent damage from abnormal electrical conditions – that spread across the interconnected grid faster than human operators could intervene or fully understand what was happening.

The 2003 blackout is relevant to this book not because a transformer directly caused it, but because it illustrates a structural characteristic of the interconnected grid architecture built up over the previous century, examined throughout Part Two: the same interconnection that provides enormous reliability benefits under normal conditions – allowing utilities to share reserve capacity and support one another during localized problems – can also, under specific cascading failure conditions, propagate a problem far beyond its point of origin, faster than automatic protective systems and human operators can always contain. Transformers, as some of the most critical and most electrically stressed pieces of equipment at any substation, are frequently among the first components that automatic protective relays disconnect when a cascading disturbance begins, a protective action that, while necessary to prevent damage to the transformer itself, can also contribute to the spread of an outage across a wider area if not carefully coordinated with the rest of the grid's protective systems.

The 2003 event led to significant reforms in how the North American grid's reliability is regulated and enforced, including the creation of mandatory, federally enforceable reliability standards administered by the North American Electric Reliability Corporation, a regulatory evolution examined in greater detail in *A History of the United States Electrical Grid.* Those reforms have meaningfully reduced the likelihood of a similarly scaled cascading failure recurring, though grid operators and regulators continue to study and refine protective coordination practices in light of the 2003 event and subsequent close calls.

The Metcalf Substation Attack

A very different kind of transformer vulnerability came into sharp public focus on April 16, 2013, when unknown attackers used rifles to fire more than one hundred rounds into Pacific Gas and Electric's Metcalf transmission substation near San Jose, California, damaging seventeen large transformers badly enough to require them to be taken out of service, in an attack sophisticated and deliberate enough that federal investigators and independent security experts have generally concluded it was carried out by a well-planned, professionally executed team rather than random vandalism, though no one has ever been publicly identified or charged in connection with the incident.

The Metcalf attack did not cause a widespread blackout – grid operators were able to reroute power and avoid customer outages, aided by the substation's connection to the broader interconnected grid architecture discussed throughout this book – but it caused tens of millions of dollars in equipment damage and, more significantly for long-term grid security planning, it demonstrated with uncomfortable clarity that a small, low-cost, low-technology attack using publicly available rifles and readily obtainable information about substation locations could disable a significant number of large transformers simultaneously, equipment that would take years, not days, to replace through the normal manufacturing supply chain discussed in Part Four.

Former Federal Energy Regulatory Commission chairman Jon Wellinghoff, among others, described the Metcalf attack afterward as "the most significant incident of domestic terrorism involving the grid that has ever occurred," a characterization that reflected less the immediate operational consequences – which were, again, successfully contained without customer outages – than the attack's demonstration of a previously under-appreciated vulnerability: a small number of coordinated, low-cost attacks on a handful of critical substations, executed simultaneously across multiple locations, could plausibly disable a meaningful fraction of the large transformer fleet serving a major metropolitan region, with recovery constrained not by any lack of utility repair expertise, but by the multi-year manufacturing lead times for replacement transformers.

Physical Security Reforms

The Metcalf attack prompted the electric utility industry, working with federal regulators, to develop new mandatory physical security standards for critical transmission substations, requiring utilities to assess the risk facing their most critical facilities and implement appropriate physical security measures – which can include enhanced fencing, surveillance systems, and in some cases visual barriers specifically designed to make it harder for an attacker positioned at a distance to identify and target specific pieces of critical equipment like large power transformers. These reforms represent a meaningful, if incomplete, response to the vulnerability Metcalf exposed, and they exist alongside the spare-transformer-sharing programs discussed in the previous chapter as parallel strategies for managing a risk that cannot be entirely eliminated: the concentration of enormous grid importance in a relatively small number of large, custom-manufactured, multi-year-lead-time pieces of equipment.

Extreme Weather and Aging Equipment

Beyond deliberate attack, transformers face failure risk from extreme weather – including the kind of severe winter storms examined at length in *The Reliability Frontier*'s discussion of the February 2021 Winter Storm Uri crisis in

Texas, during which extreme cold stressed electrical equipment across the state's grid – and from the ordinary, gradual aging process that affects any piece of equipment operating continuously for decades. This aging-fleet dimension of transformer risk, distinct from both cascading-failure and deliberate-attack scenarios, has become an increasingly urgent concern for utility planners in the 2020s, precisely because so much of the American transformer fleet was installed during the large postwar buildout described in Chapter 10, and is now approaching or exceeding the design lifespan those mid-twentieth-century engineers anticipated when they built it.

Understanding exactly how old that fleet has become, and what utilities are doing – and struggling to do – about replacing it, is the subject of Part Four, the section of this book that shifts from history and engineering to the urgent, unfolding crisis of the 2020s: an aging transformer fleet, a surging and unprecedented wave of new electricity demand, and a domestic and global manufacturing base struggling to keep pace with both at once.

Cyber Vulnerabilities Alongside Physical Ones

Beyond the physical attack vulnerability the Metcalf incident exposed, grid security planners have increasingly focused attention on cyber vulnerabilities affecting the digital monitoring and control systems, discussed in Chapters 11 and 20, that modern transformers and substations increasingly incorporate. A transformer's protective relay system, which automatically disconnects the unit under abnormal electrical conditions as described earlier in this chapter, is itself a digital, networked piece of equipment in most modern substations, and security researchers and federal agencies have documented realistic scenarios in which a sophisticated cyberattack could manipulate these protective systems, potentially causing unnecessary equipment trips or, in a more severe scenario, preventing protective equipment from operating correctly when it is genuinely needed, either of which could contribute to equipment damage or a cascading outage of the kind described earlier in this chapter's discussion of the 2003 Northeast blackout.

The electric utility industry has responded to these cyber concerns through mandatory cybersecurity standards, also administered through the North American Electric Reliability Corporation's compliance framework, requiring utilities to implement specific technical and procedural safeguards protecting the digital systems controlling critical grid equipment, including large power transformers. These standards continue to evolve as the digital sophistication of grid equipment increases and as the broader cybersecurity threat landscape facing critical infrastructure continues to develop, a dynamic, ongoing challenge rather than a problem with any permanent, final solution.

Insurance and Financial Risk Management

Utilities also manage transformer failure risk through specialized insurance arrangements and, in some cases, industry-wide mutual insurance structures specifically designed to cover the substantial replacement cost and lost-revenue consequences of a major transformer failure. Given the multi-year replacement lead times described throughout Part Four of this book, the financial exposure a utility faces from losing a critical large power transformer has grown substantially in recent years, since the cost of an extended outage or the need to lease expensive temporary mobile substation equipment while awaiting a permanent replacement can far exceed the transformer's own replacement cost. This growing financial exposure has itself become an additional argument utilities and regulators cite in support of the spare-equipment sharing programs and expanded domestic manufacturing capacity discussed in Chapters 13 and 19, since preventing an extended outage in the first place is, in almost every realistic scenario, considerably cheaper than managing the consequences after the fact.

Mobile Substations as a Stopgap

One practical tool utilities increasingly rely on to manage the consequences of an unexpected transformer failure, pending a permanent replacement, is the mobile substation: a smaller, trailer-mounted transformer and associated switching equipment that can be transported relatively quickly to a site experiencing a transformer outage and connected temporarily to restore at least partial service while a permanent replacement makes its way through the manufacturing and delivery process described in Chapters 12 and 13. Mobile substations cannot generally match the full capacity of the large permanent units they temporarily replace, meaning customers served by a mobile substation may face some ongoing service limitations or increased outage risk during the interim period, but they represent a valuable, increasingly important tool for reducing the real-world consequences of the extended replacement timelines this book's final section examines in detail.

Geomagnetic Disturbances: An Unusual Risk

A less commonly discussed but genuinely significant transformer vulnerability involves geomagnetic disturbances caused by severe solar storms, which can induce unusual, slowly varying currents – called geomagnetically induced currents – in long transmission lines and, through them, in the transformers connected to those lines. These currents differ fundamentally from the normal sixty-cycle alternating current transformers are designed to handle, and can push a transformer's magnetic core into a saturated operating state that produces excessive heating and, in a sufficiently severe event, can cause serious damage to affected units. The most frequently cited historical example is the March 1989 geomagnetic storm that caused a nine-hour blackout across the Canadian province of Quebec, along with documented transformer damage at affected utilities in the United States and elsewhere.

While severe geomagnetic storms of this scale are rare, the North American Electric Reliability Corporation has, in the years since, developed mandatory reliability standards requiring utilities to assess their transmission systems' vulnerability to geomagnetically induced currents and implement appropriate protective measures, including specialized blocking devices that can be installed to limit these unusual currents from reaching vulnerable transformers during a severe solar storm. This represents one more category of risk, alongside the cascading-failure, physical-attack, and cyberattack risks discussed elsewhere in this chapter, that grid planners must weigh when assessing the overall resilience of a transformer fleet that, as Chapter 15 describes, is aging in ways that may reduce its margin of safety against any of these various stress scenarios.

Learning From Near Misses

Beyond the well-documented incidents discussed in this chapter, the utility industry maintains extensive internal reporting and analysis processes for near-miss events – situations in which a transformer or other critical equipment came close to failure or was exposed to unusually severe stress without actually failing – recognizing that these near misses often contain valuable lessons about emerging vulnerabilities before they manifest as an actual, damaging failure. This kind of systematic near-miss analysis, common across many high-reliability industries including commercial aviation, has become an increasingly important part of how utilities manage transformer fleet risk, complementing the condition-monitoring and dissolved-gas-analysis techniques described in Chapter 11 with a broader, more systematic approach to identifying and addressing vulnerabilities before they produce an actual, customer-affecting failure.

Wildfire Risk and Transformer-Related Ignition

A further failure-related risk that has drawn increasing attention, particularly in the American West, involves electrical equipment, including transformers, as a source of wildfire ignition during dry, high-wind conditions. Utilities in California and other wildfire-prone states have implemented public safety power shutoff programs, proactively de-energizing lines and associated equipment during periods of extreme fire weather risk, a practice that trades planned, temporary outages for reduced wildfire ignition risk, and that has itself become a significant reliability and public policy issue, discussed at greater length in *The Reliability Frontier*'s treatment of California grid reliability challenges. While transformer failures are only one of several possible equipment-related wildfire ignition sources – downed conductors and failed connectors are more commonly cited causes in post-fire investigations – the broader equipment-hardening programs utilities have pursued in response to wildfire risk, including more robust transformer mounting and protective equipment in high-risk areas, represent one more category of proactive investment competing for utility capital and, indirectly, for the constrained transformer manufacturing capacity discussed throughout this book's final section.

A Framework for Thinking About Transformer Risk Broadly

The various failure modes discussed throughout this chapter – cascading system disturbances, deliberate physical attack, cyberattack, geomagnetic disturbance, and ordinary aging-related degradation – differ enormously in their likelihood and their typical consequences, but they share a common structural vulnerability this book has emphasized throughout: because large power transformers take so long to manufacture and deliver, any of these failure modes carries consequences considerably more severe today than the same failure would have carried during the more abundant manufacturing-capacity era described in Chapter 10. This shared vulnerability is precisely why utility risk managers, grid security planners, and federal policymakers increasingly discuss transformer failure risk and transformer manufacturing capacity as two sides of the same underlying resilience question, rather than as separate, unrelated concerns – a framing this book's final section adopts explicitly in examining both the current bottleneck and the industry's response to it.

PART IV – THE BOTTLENECK

Chapter 15: The Transformers We Inherited

The American electric grid entered the 2020s carrying a transformer fleet substantially built during the postwar boom described in Chapter 10 – decades in which utilities, confident in an abundant and reliable domestic manufacturing base, installed hundreds of thousands of transformers, both large power transformers at substations and power plants and smaller distribution transformers on poles and pads across every American neighborhood, generally designed and expected to last thirty to forty years under normal operating conditions, with many well-maintained units continuing to perform reliably well beyond that nominal design life.

An Aging Fleet, Quantified

Industry surveys and federal assessments conducted over the past decade have converged on a sobering set of figures. The average age of large power transformers in the United States is now widely estimated at somewhere between thirty-five and forty years, with a meaningful fraction of the fleet – commonly cited estimates run from roughly ten to fifteen percent of large power transformers currently in service – already operating beyond their originally intended design life of forty years. The Department of Energy, in assessments conducted over the past decade examining the resilience and security of the large power transformer fleet, has repeatedly flagged this aging profile as a significant vulnerability, not because an aging transformer necessarily fails on any predictable schedule – many well-maintained units, as noted above, continue operating safely for decades beyond their nominal design life – but because the statistical probability of failure due to insulation degradation, mechanical fatigue, or an underlying manufacturing defect finally manifesting itself, rises as a fleet ages collectively, even when any individual unit's remaining useful life remains difficult to predict with precision.

This aging profile did not emerge from neglect exactly, but from a combination of factors this book's companion volume, *A History of the United States Electrical Grid*, examined in detail from a regulatory and financial perspective: the deregulation era beginning in the 1990s and the market pressures it introduced, discussed at length in that book's Chapters 14 through 18, created financial incentives for many utilities – particularly those operating in competitive wholesale markets rather than traditional vertically integrated, cost-of-service regulated monopolies – to defer major capital replacement projects, including proactive transformer replacement, in favor of extending the operating life of existing equipment for as long as it could be safely and legally justified. A transformer that continues to pass its periodic inspections and oil testing, even as it approaches or exceeds its original design life, represents, from a purely short-term financial perspective, a depreciated asset requiring no immediate capital outlay, and utilities operating under cost pressures – whether from competitive market structures, rate-case scrutiny from regulators wary of approving large capital expenditures that raise customer rates, or simply conservative capital budgeting during periods of otherwise flat electricity demand growth – have had real, if not always well-publicized, incentives to defer replacement as long as prudently possible.

Flat Demand, Deferred Investment

Compounding this dynamic, American electricity demand growth was, for roughly two decades between the mid-1990s and the early 2020s, remarkably flat by historical standards, a period *The Reliability Frontier* examines in its own opening chapters as the calm before the current storm of AI-driven demand growth. Utilities planning capital investment during this long period of flat demand had little urgent incentive to expand transformer capacity beyond replacing outright failures, since existing infrastructure, however aging, was generally adequate to serve a customer base whose aggregate electricity consumption was not growing significantly year over year, thanks in part to efficiency gains in lighting, appliances, and industrial processes that offset population and economic growth.

This combination – an aging fleet, financial incentives favoring deferred replacement, and two decades of flat demand providing little urgent pressure to expand capacity – left the American transformer manufacturing industry, examined in detail in the next few chapters, with correspondingly little incentive to expand its own domestic production capacity during this same period. Why would a transformer manufacturer invest in expensive new factory capacity to serve a domestic utility customer base that was, in aggregate, ordering roughly the same volume of new equipment year after year, primarily to replace outright failures and serve modest, geographically uneven pockets of new load growth, rather than any broad-based surge in demand? The rational economic answer, for most of this period, was that it would not, and the domestic large power transformer manufacturing base accordingly contracted, consolidated, and in some cases relocated production overseas, where labor and, in some cases, regulatory compliance costs were lower, a trend examined in detail in Chapter 18.

The Arrival of Simultaneous Demand Shocks

This entire equilibrium – an aging but adequate fleet, served by a shrunken but sufficient manufacturing base, calibrated to decades of flat demand – has been overturned within a remarkably short span of years by the simultaneous arrival of several distinct but mutually reinforcing sources of new electricity demand, each examined individually in the next chapter: the rapid adoption of electric vehicles, requiring both home and public charging infrastructure that places new, geographically concentrated demands on local distribution transformers; a wave of reshored domestic manufacturing, particularly in semiconductor fabrication and battery production, driven partly by federal industrial policy and partly by supply-chain lessons learned during the COVID-19 pandemic; increasingly widespread adoption of electric heat pumps for home heating, driven by both economic and environmental considerations; and, most dramatically and most suddenly, the extraordinary electricity appetite of artificial intelligence datacenters, a phenomenon that barely existed as a meaningful grid-planning consideration before roughly 2022 and that, within just a few years, became the single largest new source of electricity demand utilities across the country were being asked to plan for.

A Collision, Not a Coincidence

It would be a mistake to treat the resulting supply-chain crisis as simply an unfortunate coincidence of timing – an aging fleet happening to reach a critical juncture just as demand happened to surge. The more accurate framing, and the one this book adopts, is that decades of underinvestment in both fleet renewal and manufacturing capacity, rational as those decisions may have appeared at the time given flat demand and competitive market pressures, left the system with essentially no slack, no reserve capacity, and no manufacturing surge capability to absorb a demand shock of the scale and suddenness the AI datacenter boom has produced. A system that had continued modest, proactive fleet renewal and maintained more robust domestic manufacturing capacity throughout the 2000s and 2010s, even at some ongoing cost premium, would have entered the 2020s considerably better positioned to absorb the current surge in demand than the system utilities and manufacturers actually inherited.

This is not offered as retrospective blame directed at any particular utility executive, regulator, or manufacturer; the decisions that produced today's tight equilibrium were, in most cases, individually reasonable responses to the specific incentives and market conditions each decision-maker faced at the time. But understanding how the system arrived at its current state of vulnerability is essential to understanding why the crisis examined in the remaining chapters of this book – a crisis of four-year lead times, soaring equipment costs, and interconnection queues stretching years into the future – cannot be resolved quickly or easily, no matter how much political or executive urgency is eventually brought to bear on it. The next chapter turns to the demand side of this collision in detail, examining exactly how much new electricity demand is arriving, from which sources, and why artificial intelligence datacenters in particular have upended utility planning assumptions so dramatically and so suddenly.

How Utilities Learned the Scale of the Problem

Much of the detailed data on transformer fleet age now cited throughout the utility industry and by federal agencies emerged from a series of assessments conducted over the past fifteen years, as the Department of Energy and industry organizations began systematically surveying utilities about their large power transformer inventories in response to growing concern about both aging infrastructure and the physical and cyber vulnerabilities discussed in Chapter 14. These surveys revealed not just the aggregate average age figures cited earlier in this chapter, but significant variation across regions and utility types, with some older, more established utility service territories in the Northeast and Midwest carrying noticeably older average fleet ages than faster-growing regions in the Southeast and West, where more recent population and industrial growth has driven more continuous transformer replacement and expansion simply to keep pace with new customer connections.

Depreciation Schedules and the Economics of Deferral

It is worth explaining briefly, for readers unfamiliar with utility accounting practices, exactly how the financial incentive toward deferred transformer replacement described in this chapter actually operates. Utilities typically depreciate major capital equipment, including transformers, over an assumed useful life for accounting and rate-setting purposes, often forty years or more for large power transformers. Once a transformer is substantially or fully depreciated on a utility's books, it represents, from a pure balance-sheet perspective, an asset that continues providing service at essentially no ongoing capital cost, aside from routine maintenance – a dynamic that can create a rational, if shortsighted, incentive to continue operating an aging but still-functional unit rather than proactively replacing it with a new unit that would require fresh capital investment and, in regulated utility contexts, a rate case proceeding to recover that investment from customers, a process that can be lengthy, contentious, and uncertain in its outcome.

This is not to suggest utility executives have acted recklessly or irresponsibly; regulators, for their part, have historically scrutinized utility capital spending requests carefully, mindful of the direct impact on customer electricity rates, and utilities proposing large proactive equipment replacement programs before existing equipment has actually failed or shown clear signs of imminent failure have sometimes faced regulatory skepticism about whether such spending is truly necessary or prudent. The result, across the industry as a whole, has been a systematic bias toward reactive rather than proactive transformer replacement – fixing or replacing units after clear signs of degradation appear, rather than replacing them preemptively based on age alone – a bias that made sense under the flat-demand conditions of the 1990s through early 2020s but that has left the industry with less replacement runway than it might otherwise have had entering the current demand surge.

The Federal Data Picture

The Department of Energy's periodic large power transformer assessments, along with data collected by the North American Electric Reliability Corporation and various utility industry trade associations, have converged on estimates suggesting the median large power transformer currently in service was manufactured sometime in the 1980s, with meaningful numbers of units dating to the 1960s and 1970s still in active service across various utility territories, particularly in older, established metropolitan areas whose grid infrastructure was largely built out during the postwar expansion described in Chapter 10. These federal assessments have increasingly framed transformer fleet age not merely as an operational maintenance concern but as a matter of genuine national infrastructure and economic security, given the central role transformers play in virtually every aspect of the demand growth story examined in the following chapter.

Distribution Transformers Face Their Own Aging Crisis

While this chapter and much of Part Four's discussion has focused primarily on large power transformers, given their outsized role in the AI datacenter and reshored-manufacturing interconnection stories examined in subsequent chapters, it would be a mistake to overlook the parallel aging crisis affecting the much larger population of smaller distribution transformers serving individual neighborhoods and businesses across the country. Many of these units, as discussed in Chapter 9, were installed during the rural electrification and postwar expansion eras and have remained in service for fifty years or more, and utilities report that distribution transformer lead times, while generally shorter in absolute terms than large power transformer lead times, have nonetheless stretched considerably compared to historical norms, complicating the routine, unglamorous work of replacing aging neighborhood equipment before it fails and causes localized outages.

This distribution-level dimension of the crisis receives less public and media attention than the dramatic, headline-grabbing story of AI datacenters waiting years for grid interconnection, but it arguably affects a broader cross-section of ordinary utility customers more directly, since a failed neighborhood distribution transformer, however routine an event historically, now carries a meaningfully greater risk of extended outage while a replacement unit works its way through a supply chain considerably more strained than the one utilities relied upon as recently as the late 2010s.

Utility Fleet Management Strategies

In response to these combined pressures, utilities have increasingly adopted more sophisticated fleet management strategies, using the condition-monitoring techniques described in Chapter 11 to prioritize which aging units most urgently need replacement, maintaining larger strategic spare-parts inventories where feasible, and, in some cases, extending planned maintenance intervals or performing more extensive refurbishment on aging units to extend their safe service life further than utilities might have found prudent or necessary during the more abundant equipment-availability era described in Chapter 10. These strategies represent a reasonable, pragmatic adaptation to genuinely difficult circumstances, but they also illustrate, once again, a theme running throughout this book's final section: the transformer bottleneck is not a problem utilities can simply purchase their way out of on a short timeline; it is a structural constraint that requires careful, often uncomfortable prioritization of scarce equipment and scarce capital across genuinely competing needs, a challenge the next several chapters examine from multiple angles.

Regulatory Efforts to Encourage Proactive Replacement

In recognition of the deferred-replacement dynamic described in this chapter, some state utility regulators have begun exploring reforms intended to reduce the disincentive toward proactive transformer replacement, including mechanisms that allow utilities to recover the cost of proactive, risk-based equipment replacement programs more predictably than the traditional, largely reactive rate-case process historically allowed. These reforms remain unevenly adopted across different states, reflecting the broader patchwork of utility regulatory approaches that *A History of the United States Electrical Grid* examines in detail, and their ultimate effectiveness at meaningfully accelerating fleet renewal, given the manufacturing capacity constraints discussed throughout Part Four, remains an open question regulators and utilities are actively working through.

The Insurance Value of Accurate Condition Data

The condition-monitoring techniques discussed in Chapter 11 have taken on heightened strategic importance precisely because of the aging-fleet dynamics described in this chapter: with replacement lead times stretched to years, utilities can no longer simply plan to replace a transformer reactively once it shows serious signs of degradation, since ordering a replacement at that point may leave a utility managing a genuinely elevated failure risk for years while awaiting delivery. Instead, utilities increasingly need to identify units showing early, subtle signs of degradation years before those units would otherwise be flagged for replacement under traditional inspection practices, effectively front-loading the equipment-ordering decision to compensate for the extended manufacturing lead time now standard across the industry. This shift represents a genuinely significant change in how utilities approach fleet management, discussed further in Chapter 20's treatment of predictive maintenance technology, and it illustrates one more way the manufacturing bottleneck examined throughout this book's final section has reshaped utility operating practice, well beyond its most visible effects on new customer interconnection timelines.

Chapter 16: Everything Wants Electricity

For roughly two decades, as described in the previous chapter, American utility planners built their long-term capital plans around demand forecasts that assumed, reasonably given the historical pattern, that aggregate electricity consumption would grow slowly if at all, with efficiency gains largely offsetting population and economic growth. That assumption has collapsed within the span of just a few years, replaced by demand forecasts that, in many regions, project growth rates utilities have not seen since the postwar boom chronicled in Chapter 10 – and this chapter examines, source by source, exactly where that new demand is coming from, and why it converges, more than most people realize, on the exact same category of scarce equipment: large power transformers.

Electric Vehicles

Electric vehicle adoption, while its pace has varied year to year and remains a subject of active political and market debate, has grown from a negligible share of new vehicle sales a decade ago to a meaningful and steadily increasing share today, and every electric vehicle charging at home, at a workplace, or at a public charging station draws power through the same distribution transformer infrastructure serving every other customer on that local circuit. A single home Level 2 charger can draw as much continuous power as several traditional major appliances combined, and a cluster of homes on the same neighborhood transformer all installing home chargers within a short period – a pattern utilities have observed repeatedly as EV adoption spreads unevenly through specific neighborhoods and communities – can push a decades-old distribution transformer, sized originally for a very different mix of household appliances, beyond its safe rated capacity, requiring utilities to proactively upgrade or replace transformers in neighborhoods experiencing concentrated EV adoption well before those transformers would otherwise have reached the end of their service life.

Public and fleet charging infrastructure presents an even more concentrated version of this same challenge. A fast-charging station designed to serve multiple vehicles simultaneously with high-power direct-current charging can require electrical service comparable to a small industrial facility, often necessitating entirely new, dedicated distribution transformers or even new substation capacity in locations – along highway corridors, for instance – that may never have hosted a comparably large single electrical load before.

Electrification of Heating and Reshored Manufacturing

Running parallel to vehicle electrification is a broader trend toward electrifying home heating, primarily through heat pumps, which are more energy-efficient than the fossil-fuel furnaces they replace but which, particularly in regions with cold winters, can substantially increase a home's peak electricity demand during the coldest periods of the year – precisely the same winter peak-demand periods that already strain grid reliability, as *The Reliability Frontier* examined at length in its discussion of Winter Storm Uri and subsequent cold-weather grid emergencies. A home switching from gas heating to an electric heat pump, multiplied across an entire neighborhood or region, can meaningfully increase the peak load a local distribution transformer must serve during exactly the season when transformer performance is already most stressed by cold-weather conditions.

Simultaneously, a wave of new domestic manufacturing investment – semiconductor fabrication plants, battery manufacturing facilities, and other advanced manufacturing projects, spurred by federal industrial policy initiatives adopted in the early 2020s alongside supply-chain security concerns highlighted by the COVID-19 pandemic – has brought a series of large, electricity-intensive industrial facilities online or under construction in specific regions, each requiring dedicated, often custom-engineered large power transformers to serve loads that, in some cases, rival the electricity consumption of a small city, concentrated at a single industrial site.

The AI Datacenter Surge

None of the demand sources described above, however, has upended utility planning assumptions as dramatically, or as suddenly, as the electricity appetite of artificial intelligence datacenters, the central subject of my previous book, *The Reliability Frontier*. That book examined this phenomenon primarily from the perspective of grid operations, capacity markets, and utility planning; this book adds the specific engineering dimension that book could only gesture toward: every one of the massive datacenter interconnection requests utilities have received since roughly 2022 requires, near its point of connection to the grid, one or more large power transformers, frequently at extra-high voltage, custom-engineered to the specific electrical characteristics of that datacenter's connection point and often representing a meaningfully large fraction of the total large power transformer manufacturing capacity available to serve an entire utility's broader service territory.

A single hyperscale AI training campus can request interconnection service comparable to that of a mid-sized American city – several hundred megawatts, and in some proposed projects, over a gigawatt of capacity, an amount of power that, a decade ago, would have been associated only with an entire metropolitan area's demand, not a single industrial customer. Serving a load of this magnitude typically requires not just generation capacity, a subject *The Reliability Frontier* examined extensively, but a dedicated set of large power transformers stepping transmission voltage down to whatever voltage the datacenter's own internal electrical systems require – transformers that must be specifically engineered, ordered, manufactured, tested, and delivered before the datacenter can actually draw the power its developers have contracted for, regardless of how quickly the datacenter's own construction otherwise proceeds.

Why AI Demand Is Different in Kind, Not Just Degree

It is worth being precise about what makes the AI datacenter demand surge different from the other demand sources described in this chapter, beyond its sheer scale. Electric vehicle adoption, heat pump adoption, and reshored manufacturing investment, while individually significant, have generally grown at a pace that, however challenging, at least resembles historical patterns of gradual, multi-year demand growth that utility planning processes were originally designed to accommodate. The AI datacenter demand surge, by contrast, has arrived with a speed and a per-project scale that overwhelms traditional utility interconnection planning timelines, which were built around an assumption that new large industrial customers would arrive occasionally, one at a time, giving utilities years of advance notice to plan and build the necessary transmission and transformer infrastructure.

Instead, utilities across the country have found themselves receiving interconnection requests from multiple hyperscale datacenter developers simultaneously, each requesting service on a timeline measured in a small number of years, sometimes competing for interconnection capacity within the same specific transmission corridor or substation service territory, and each request, individually, potentially requiring large power transformer capacity that a utility might previously have expected to order gradually over a much longer period to serve a more diverse and more gradually growing set of customers. This concentrated, simultaneous, multi-gigawatt scale of new demand is precisely what has transformed the large power transformer from an important but generally manageable procurement item into the specific, named bottleneck utility executives, industry analysts, and increasingly financial journalists now identify as the single greatest constraint on how quickly new electricity demand – AI datacenters most visibly, but also reshored manufacturing, vehicle electrification, and heating electrification – can actually be served.

Setting Up the Central Question

The remainder of this book's final section addresses the question this convergence of demand sources inevitably raises: why, specifically, can the transformer manufacturing industry not simply expand production to meet this surge in demand, the way many other industries have historically responded to sudden demand shocks by building new factory capacity or expanding existing production lines? The next chapter examines the concrete, measurable consequences of this mismatch between demand and supply – the lead times, backlogs, and price increases utilities are actually experiencing in the current market – before subsequent chapters turn to the underlying manufacturing and supply-chain constraints that explain why those consequences have proven so severe and so difficult to resolve quickly.

Quantifying the AI Datacenter Demand Wave

Industry and utility forecasts published throughout the mid-2020s have attempted to quantify just how much new electricity demand the AI datacenter buildout represents, with estimates varying considerably depending on methodology and assumptions about how much of the publicly announced datacenter development pipeline will actually be built, but generally converging on the conclusion that datacenter electricity demand, which represented a small single-digit percentage of total U.S. electricity consumption as recently as the early 2020s, could plausibly grow to represent a meaningfully larger share – estimates commonly range from high single digits to low double digits as a percentage of total national electricity consumption – within just several years, an extraordinarily rapid demand shift for an electricity system whose infrastructure, as described throughout this book, typically takes years to expand in response to new demand.

Regional concentration compounds this challenge considerably: datacenter development has clustered heavily in specific regions offering some combination of available land, favorable climate for cooling, existing fiber-optic network infrastructure, and, critically, some prospect of available electric grid capacity, including Northern Virginia's long-established "Data Center Alley," parts of Texas within the ERCOT grid, and a growing number of additional regions across the Southeast, Midwest, and Mountain West as developers seek out locations with available interconnection capacity. This geographic concentration means the transformer and transmission bottleneck described throughout Part Four of this book is not evenly distributed nationally; certain specific utility service territories and transmission corridors face vastly more acute pressure than others, a pattern *The Reliability Frontier* examined in detail when discussing region-specific interconnection queue backlogs.

Behind-the-Meter Generation as a Partial Workaround

The scale and urgency of AI datacenter electricity demand has prompted a growing number of developers to pursue behind-the-meter or co-located generation arrangements, in which the datacenter's electricity is supplied substantially or entirely by generation located on or very near the same site, rather than relying primarily on transmission-connected utility service. These arrangements, which can involve natural gas generation, and in a growing number of proposed projects, dedicated small modular nuclear reactors or large-scale battery storage paired with renewable generation, can reduce a datacenter's dependence on new transmission-connected transformer capacity, though rarely eliminate it entirely, since even a substantially self-sufficient datacenter typically maintains some grid interconnection for backup reliability and to sell or purchase power as its own generation and demand fluctuate.

This strategy, examined further in Chapter 19, illustrates a broader theme running through this book's final section: when the traditional utility-transformer supply chain cannot deliver capacity on the timeline large customers require, those customers with sufficient capital increasingly seek alternative paths that partially route around the bottleneck, a market response that, while rational for individual developers, raises the broader questions about equity and grid-planning coordination that Chapter 19 examines in detail.

Comparing Demand Sources Side by Side

It is useful to step back and compare the demand sources discussed in this chapter directly, because they differ not only in scale but in the specific kind of transformer capacity they strain. Electric vehicle charging and heat pump adoption primarily strain distribution-level transformers – the smaller units serving individual neighborhoods and residential circuits – because their effects, while collectively enormous, arrive customer by customer, incrementally, across a utility's entire residential service territory. Reshored manufacturing facilities and AI datacenters, by contrast, primarily strain large power transformers and transmission-level infrastructure, because a single project can represent as much incremental load as an entire town, concentrated at one interconnection point.

This distinction matters for how utilities and manufacturers should think about capacity planning, because it means the current bottleneck is not a single, uniform problem with a single solution; it is really two related but distinct bottlenecks operating simultaneously, one in distribution transformer manufacturing and one in large power transformer manufacturing, each drawing on an overlapping but not identical set of raw materials, factories, and skilled labor. Some manufacturers specialize primarily in one category or the other, meaning capacity expansion in one segment does not automatically relieve pressure in the other, a nuance that sometimes gets lost in general media coverage that treats "the transformer shortage" as a single undifferentiated phenomenon.

A Personal Illustration

It may help to make this demand convergence tangible through a single composite, illustrative scenario, drawn from patterns documented across multiple utility service territories rather than any single specific case. Imagine a mid-sized utility serving a metropolitan area that, as recently as 2018, was planning for electricity demand growth of perhaps one percent annually, driven by modest population growth and new commercial construction. By 2024, that same utility finds itself simultaneously fielding interconnection requests from: a proposed AI datacenter campus requesting several hundred megawatts of capacity; a battery manufacturing facility relocating production to the United States and requesting a dedicated large power transformer of its own; a rapidly growing EV charging network requiring dozens of new or upgraded distribution transformers across its service territory; and its own aging fleet of 1970s- and 1980s-vintage substation transformers, several of which condition-monitoring data suggests are approaching the end of their safe operating life. Each of these needs, individually, would have been manageable under the utility's historical planning assumptions. Together, arriving within the same few years, they exceed what the utility's existing relationships with transformer manufacturers can supply on any timeline faster than the three-to-five-year lead times described in the next chapter – precisely the convergence this chapter has described in more general terms.

Water, Cooling, and the Datacenter's Other Infrastructure Demand

It is worth noting, briefly, that electricity is not the only scarce infrastructure resource AI datacenters compete for, even though this book's focus remains squarely on the transformer bottleneck. Datacenter cooling systems, particularly for the densest, most power-intensive AI training hardware, increasingly require substantial water resources in regions using evaporative cooling technology, a parallel resource-competition story that has drawn its own share of local political controversy in water-stressed regions where large datacenter projects have been proposed. This water dimension does not directly involve transformers, but it illustrates a broader pattern worth naming: the AI datacenter buildout is straining multiple categories of physical infrastructure simultaneously, electricity and water prominent among them, and utilities and local governments evaluating a proposed datacenter project increasingly must weigh transformer and transmission capacity constraints alongside water availability, land use, and community relations concerns in ways that add further complexity to the already difficult interconnection planning process described throughout this chapter and the next.

Datacenter Siting as a Response to the Bottleneck

Recognizing the severity of the transformer and transmission bottleneck in some of the most heavily developed datacenter regions, including Northern Virginia, some developers have begun shifting new project proposals toward regions with more readily available grid interconnection capacity, even where those regions may otherwise be somewhat less attractive from a fiber connectivity or labor market perspective. This siting shift represents a market response to the bottleneck described throughout this book's final section, functioning similarly to how manufacturers historically located near sources of abundant, low-cost power, as Niagara Falls-area industry did in the decades following the events described in Chapter 6. Regions with newer, less heavily loaded grid infrastructure, or with substantial existing generation capacity relative to local demand, have in some cases seen a meaningful uptick in datacenter development interest specifically because they offer a faster path to reliable interconnection than more saturated markets, a dynamic that may, over time, help distribute the demand surge described in this chapter more evenly across the national grid, easing pressure in the most severely constrained specific locations even as aggregate national demand continues its rapid growth.

A Regulatory Reckoning With Load Forecasting

The demand convergence described throughout this chapter has also forced a broader reckoning within the utility planning profession itself, since the load-forecasting methodologies utilities have relied upon for decades were generally built around the assumption of gradual, predictable, broadly diversified demand growth across many small and medium customers, an assumption the AI datacenter phenomenon has upended given how much aggregate load growth can now originate from a comparatively small number of very large, very fast-moving individual projects. Utility planners and the regulators who review their forecasts have increasingly had to develop new methodologies specifically for evaluating the credibility and likely timeline of large, individually announced datacenter projects, recognizing that not every announced project will ultimately be built as originally proposed, while also recognizing the real risk of underestimating a demand wave that has, in most regions, exceeded even the more aggressive forecasts utilities published just a few years earlier. Getting this forecasting exercise right matters enormously for the transformer procurement decisions discussed throughout this book's final section, since ordering too conservatively risks perpetuating the bottleneck, while ordering too aggressively risks stranding costly, long-lead-time equipment if specific projects fail to materialize as forecast – a genuinely difficult planning problem with no easy resolution, and one more dimension of the broader challenge this book's remaining chapters examine from the supply side of the equation.

Chapter 17: The Four-Year Wait

The consequences of the demand surge described in the previous chapter, colliding with the aging fleet and contracted manufacturing base described in Chapter 15, are no longer a matter of industry speculation or forward-looking projection; they are a documented, quantifiable, present-tense reality that utility executives, industry trade groups, and federal agencies have been tracking with increasing alarm throughout the mid-2020s. This chapter lays out exactly what that reality looks like, in the plainest terms available.

Lead Times That Would Have Seemed Absurd a Decade Ago

As recently as the early 2010s, a utility ordering a standard large power transformer from a domestic or international manufacturer could typically expect delivery within six months to a year, depending on the specific voltage class and customization required, a timeline consistent with the manufacturing maturity described in Chapter 10. By the middle of the 2020s, industry surveys and direct utility reporting have documented lead times for large power transformers commonly running three to five years from initial order to delivery, with the most specialized, highest-voltage units, or those ordered during periods of particularly acute backlog, sometimes quoted even longer. Smaller, though still substantial, distribution transformers – the kind serving individual neighborhoods and smaller commercial and industrial customers – have seen lead times stretch from historical norms measured in weeks or a small number of months to, in many cases, a year or more, a disruption that, while less dramatic in absolute terms than the large power transformer story, has nonetheless meaningfully slowed routine grid maintenance, storm restoration equipment replenishment, and ordinary economic development projects across the country.

These lead time figures are not uniform across every manufacturer, every voltage class, or every region, and utilities with long-standing manufacturer relationships, larger annual order volumes, or willingness to pay premium pricing for expedited production have in some cases secured somewhat shorter lead times than the most severe figures reported industry-wide. But the general direction and rough magnitude of the shift – from a world of six-to-twelve-month lead times to a world of three-to-five-year lead times for the largest units – is consistent across virtually every industry survey, utility earnings call, and trade publication covering the sector through the mid-2020s.

Price Increases Compounding the Delay

Lead times alone would represent a serious planning challenge for utilities, but the current transformer market has compounded that challenge with substantial price increases as well. Industry pricing data collected by trade associations and equipment brokers has documented large power transformer prices roughly doubling, and in some voltage classes and specifications more than doubling, over the first half of the 2020s compared to prices prevailing in the late 2010s, driven by the combination of raw material cost increases – particularly for the specialized electrical steel and copper discussed in the next chapter – and straightforward supply-and-demand pricing power, as a limited number of manufacturers face order backlogs stretching years into the future and can price accordingly.

For utilities, this combination of longer lead times and higher prices creates a genuinely difficult capital planning problem: a transformer ordered today to serve demand growth utilities can already see coming may not arrive for three to five years, at a price that may itself have increased further by the time of delivery, requiring utilities to commit substantial capital years in advance of actually needing the equipment, with limited ability to adjust the order's specifications or timing in response to changing circumstances once the order is placed and the manufacturer's production slot is reserved.

Interconnection Queues and the AI Datacenter Consequence

The most visible and, for the general public, most legible consequence of this transformer bottleneck is its direct impact on the interconnection queues utilities use to process requests from new large customers, including the AI datacenter developers whose demand *The Reliability Frontier* examined extensively. When a utility receives an interconnection request from a large new customer, one of the central questions its engineers must answer is when the necessary transformer and transmission infrastructure can actually be built and energized to serve that customer safely and reliably – and increasingly, the honest answer to that question is constrained less by the utility's own construction and engineering capacity than by the multi-year lead time required simply to obtain the transformers the project needs.

This has produced the specific, now widely reported phenomenon of datacenter developers being told by utilities that reliable grid interconnection, at the scale and location the developer originally proposed, will not be available for three, four, or in some documented cases even more years – a timeline that can meaningfully affect a datacenter developer's own project economics and site selection decisions, and that has, in a number of publicly reported cases, led hyperscale technology companies to explore alternative strategies entirely, including on-site or co-located generation that bypasses some (though rarely all) of the transformer and transmission bottleneck, a strategy examined further in Chapter 19.

Not Only Datacenters

It bears emphasizing, because coverage of this issue in the general press has understandably focused heavily on the dramatic scale and high public profile of AI datacenter projects, that the transformer bottleneck affects a far broader range of utility customers and projects than datacenters alone. Utilities attempting to interconnect new solar and wind generation projects – an important theme in *The Reliability Frontier*'s discussion of the renewable energy transition – have reported similar transformer-driven delays affecting their own interconnection queues, since new generation projects require their own dedicated transformers to step generator output up to transmission voltage, competing for the same constrained manufacturing capacity as datacenter loads. Reshored manufacturing facilities, discussed in the previous chapter, face comparable delays. And utilities' own routine grid maintenance and storm-hardening programs, which regularly require replacing aging transformers identified through the condition-monitoring practices described in Chapter 11, have in some documented cases been forced to defer planned replacements of aging equipment specifically because replacement units cannot be obtained on a reasonable timeline – an uncomfortable feedback loop in which the very equipment shortage examined in this chapter makes it harder to address the aging-fleet vulnerability examined in Chapter 15, potentially increasing the risk of the kind of unplanned failure examined in Chapter 14.

A Bottleneck With No Quick Fix

Perhaps the most sobering aspect of the current transformer crisis, and the aspect this book's remaining chapters examine in detail, is that there is no readily available quick fix. Unlike some supply-chain disruptions that resolve within a matter of months as manufacturers ramp up existing production capacity or redirect inventory, the transformer bottleneck is rooted in fundamental, physical manufacturing capacity constraints – specialized factories, specialized raw materials, and specialized skilled labor, each requiring years, not months, to expand meaningfully. Understanding exactly what those constraints are, and why they cannot simply be resolved by writing larger purchase orders or offering manufacturers more attractive pricing, is the subject of the next chapter, which examines the transformer factory floor itself and the specific materials and skills bottlenecked there.

How Utilities Are Reporting the Crisis Publicly

Evidence of the transformer bottleneck's severity has become increasingly visible not just in specialized trade publications but in the routine public disclosures utilities make to investors and regulators. Quarterly earnings calls held by major publicly traded utility holding companies throughout the mid-2020s have featured recurring executive commentary on transformer procurement challenges, with company leadership routinely citing extended lead times and elevated equipment costs as factors affecting capital spending plans and, in some cases, the pace at which new customer interconnections, including large datacenter and industrial loads, can be completed. State utility commission proceedings, where utilities must justify proposed capital spending and rate increases to regulators, have similarly featured extensive testimony documenting current transformer market conditions, providing a rich, if technical, public record of exactly how severe and how widespread the bottleneck has become across different utility service territories and equipment categories.

A Regional Case Study

While this book generally avoids naming specific utilities to keep its focus on the industry-wide pattern rather than any single company's particular circumstances, the broad pattern reported across multiple regions follows a consistent shape: a utility receives interconnection requests, often from multiple large customers simultaneously, that collectively exceed the transformer and transmission capacity currently available or plannable within a normal multi-year utility capital planning cycle; the utility places orders for the necessary new transformers, encountering the three-to-five-year lead times described above; and the utility communicates to affected customers, including datacenter developers, manufacturing facility operators, and in some cases entire new housing developments or municipal expansions, that full-capacity interconnection will not be available until those new transformers arrive, sometimes several years in the future, leaving customers to decide whether to wait, scale back their initial electricity requirements to fit within currently available capacity, pursue partial or interim service arrangements, or explore alternative sites or self-generation strategies of the kind discussed in the previous chapter.

The Compounding Effect on Grid Planning

Beyond its direct effect on individual interconnection timelines, the transformer bottleneck has begun to alter how utilities approach long-term grid planning more broadly, encouraging a shift toward ordering equipment further in advance of confirmed need than traditional utility planning practice previously considered prudent, essentially building in additional lead time buffer to account for the multi-year manufacturing reality described in this chapter. This shift carries its own financial and regulatory complications, since ordering equipment years before a specific project's need is confirmed exposes utilities, and ultimately their customers, to the risk of having ordered equipment for a project that is later canceled, delayed, or relocated – a genuine planning tension that regulators and utility executives are actively working through as the industry adjusts to a fundamentally longer equipment-procurement timeline than the one it operated under for most of the previous half-century.

The Used and Refurbished Equipment Market

One notable market response to the lead-time and pricing pressures described in this chapter has been growing utility interest in used, refurbished, or previously ordered-but-unclaimed transformers, a market segment that received relatively little attention before the current crisis but that has grown substantially as utilities search for any available path to obtaining equipment faster than the standard new-manufacture lead time. Specialized brokers and refurbishment firms now actively source transformers from decommissioned industrial facilities, canceled projects, or utilities with excess inventory, refurbishing and re-testing them to confirm they meet the electrical and mechanical requirements of a new installation site before reselling them, often at a substantial premium given the scarcity value the current market environment has created.

This secondary market, while genuinely useful as a partial release valve for the most acute cases – a utility facing an unplanned failure with no realistic prospect of a new unit for several years may reasonably choose a suitable refurbished unit available in months rather than waiting for new manufacture – cannot resolve the underlying supply-demand imbalance described throughout this chapter, since the total supply of available used and refurbished units is inherently limited by the size of the existing installed fleet, and cannot expand in response to demand the way new manufacturing capacity eventually can, given sufficient time and investment, as the next chapter describes.

Utility Rate Cases and the Cost Pass-Through

The substantial price increases documented in this chapter inevitably flow, in large part, into the rates utility customers pay, since utilities operating under traditional cost-of-service regulation are generally permitted to recover the cost of prudently incurred equipment purchases through the rate-setting process overseen by state utility commissions. This has made transformer procurement costs an increasingly visible line item in utility rate case proceedings throughout the mid-2020s, with consumer advocates and regulators scrutinizing utility equipment purchasing decisions more closely than in the past, given the magnitude of the price increases involved, while utilities have argued, generally successfully, that the price increases reflect genuine, well-documented market conditions rather than any imprudent purchasing decision on the utility's part. This regulatory dynamic adds yet another dimension to the transformer bottleneck's real-world consequences: not merely delayed interconnections and constrained grid expansion, but rising electricity rates for ordinary residential and commercial customers, layered atop the other cost pressures – including the renewable energy transition and grid-hardening investments discussed in *The Reliability Frontier* – already affecting utility bills across the country.

How Smaller Utilities and Cooperatives Are Coping

Much of this book's discussion of the transformer bottleneck has implicitly focused on large investor-owned utilities serving major metropolitan areas, given their prominent role in the AI datacenter interconnection stories that have dominated media coverage. Smaller municipal utilities and the rural electric cooperatives descended from the Rural Electrification Administration era described in Chapter 9 face many of the same lead-time and pricing pressures described throughout this chapter, but often with less purchasing leverage and fewer alternative options than their larger counterparts, since manufacturers, facing more orders than they can fill promptly, understandably prioritize their largest, most consistent customers when allocating scarce production capacity.

Industry associations representing these smaller utilities, including national organizations serving rural electric cooperatives and municipal utilities, have responded by developing cooperative purchasing arrangements and shared spare-equipment pools, extending the kind of mutual-aid tradition already well established in the cooperative movement's history, described in Chapter 9, to the specific challenge of transformer procurement during a period of industry-wide scarcity. These smaller-utility cooperative purchasing efforts cannot fully offset the scale advantages larger utilities bring to their own manufacturer negotiations, but they represent a meaningful, actively evolving response that helps ensure the current bottleneck's consequences, while real, do not fall disproportionately on the customers of the smallest and least resourced utility systems.

Insurance Rate Increases as a Downstream Consequence

A further downstream consequence of the extended replacement timelines and elevated equipment costs discussed throughout this chapter has been rising property and business-interruption insurance costs for utilities themselves, as insurers price in the increased financial exposure a utility faces from a major unplanned transformer failure given the multi-year replacement timeline now involved. These rising insurance costs represent one more, relatively subtle way the transformer bottleneck's consequences ripple through utility finances and, ultimately, customer rates, beyond the more direct and visible consequences – delayed interconnections and elevated equipment purchase prices – that have received the bulk of public and regulatory attention.

Comparing Today's Backlog to Historical Equipment Shortages

Utility industry veterans occasionally point out that the electrical equipment sector has faced supply constraints before, including periods during and after the Second World War described briefly in Chapter 8, and shortages of specific components during earlier decades of rapid grid expansion. What distinguishes the current transformer bottleneck from these historical precedents, most industry analysts agree, is its combination of duration, breadth, and root cause: previous shortages tended to resolve within a year or two as wartime priorities shifted or as temporary supply disruptions eased, whereas the current bottleneck stems from structural, multi-year manufacturing capacity and raw-material constraints that cannot resolve on a comparably short timeline, as Chapter 18 has explained in detail. This distinction matters for how utilities, regulators, and policymakers should calibrate their response: treating the current situation as a temporary disruption likely to resolve on its own within a year or two, as some earlier equipment shortages did, would risk underinvesting in exactly the kind of sustained, multi-year capacity-building response Chapter 19 describes as genuinely necessary.

Customer Communication and Managing Expectations

A final, more mundane but genuinely important consequence of the extended lead times described in this chapter involves the customer communication challenge utilities now face when explaining project delays to affected commercial and residential customers unfamiliar with the underlying manufacturing dynamics this book has described in detail. Utilities have increasingly invested in clearer, more proactive customer communication about realistic interconnection timelines, recognizing that customers and developers who understand the genuine, structural nature of the delay – rather than perceiving it as bureaucratic slowness or utility indifference – are generally better equipped to make informed decisions about project timing, site selection, and interim alternatives, a communication challenge this book's own effort to explain the transformer bottleneck to a general audience is, in its own modest way, intended to help address.

Chapter 18: The Factory Floor

To understand why the large power transformer industry cannot simply expand production quickly in response to the demand surge and price signals described in the previous two chapters, it helps to walk through, one by one, the specific materials, components, and skills that go into building a large power transformer, and to understand why each one represents its own distinct and difficult-to-resolve bottleneck, rather than a single simple constraint that a single policy fix or investment could quickly address.

Grain-Oriented Electrical Steel

The single most frequently cited material constraint in the current transformer manufacturing crisis is grain-oriented electrical steel, the specialized core material described in Chapter 11, whose distinctive crystal-grain alignment is what allows transformer cores to channel magnetic fields efficiently with minimal energy loss. This steel is not produced through an ordinary steel mill's standard rolling process; it requires a specialized, capital-intensive production process involving precise, repeated cold rolling and high-temperature annealing steps, carried out at a genuinely small number of specialized steel mills worldwide, with meaningful domestic production capacity concentrated in a handful of American facilities and a somewhat larger, but still limited, number of facilities in Japan, South Korea, Germany, and a small number of other countries.

Expanding grain-oriented electrical steel production capacity is not a matter of simply running an existing mill for more hours; it typically requires substantial new capital investment in specialized rolling and annealing equipment, and the construction lead time for this kind of specialized industrial capacity is itself measured in years, not months. Several producers have announced capacity expansion investments in response to the current transformer demand surge, but the physical construction, commissioning, and ramp-up of new grain-oriented electrical steel production capacity takes long enough that even aggressive, well-funded expansion announcements made in the early 2020s were not expected to meaningfully increase available supply until the second half of the decade at the earliest – illustrating precisely why this bottleneck cannot resolve on the kind of timeline that would meet the demand surge described in Chapter 16.

Copper and Its Competing Demands

Copper, used extensively in transformer windings as described in Chapter 11, faces a different but related constraint: it is not a specialty material produced by only a handful of mills, but a globally traded commodity whose price and availability are influenced by a wide range of demands entirely unrelated to transformer manufacturing, including the copper requirements of the broader electrification and renewable energy buildout that is itself part of the demand surge driving transformer orders in the first place. Electric vehicles, wind turbines, solar installations, and the broader grid infrastructure buildout required to serve all of the new demand sources discussed in Chapter 16 all compete for the same global copper supply, contributing to copper price increases that compound the cost pressures facing transformer manufacturers and, ultimately, the utilities and customers who purchase their products.

Some manufacturers have responded to copper cost and availability pressure by substituting aluminum windings in certain transformer designs where aluminum's somewhat lower electrical conductivity can be accommodated through larger conductor cross-sections, a workable engineering tradeoff for some applications but not a universal substitute, particularly for the most demanding, most compact high-voltage designs where aluminum's lower conductivity and different mechanical properties present real design challenges.

Insulating Materials

Beyond core steel and winding conductor, large power transformers require substantial quantities of specialized insulating paper, pressboard, and oil, each produced through its own specialized supply chain, discussed at length in Chapter 11's description of transformer internal construction. While generally less severely constrained than electrical steel, these insulating material supply chains have nonetheless faced their own pressures during the current demand surge, particularly for the highest-grade materials required in extra-high-voltage designs, and manufacturers have in some cases reported insulating material lead times lengthening alongside the broader transformer supply chain squeeze.

Skilled Labor: The Bottleneck That Cannot Be Automated Away

Perhaps the least appreciated, and in some respects most difficult to resolve quickly, constraint facing the transformer manufacturing industry is skilled labor. Winding the coils for a large power transformer, particularly the largest, most demanding extra-high-voltage designs, remains, even in the 2020s, a process that depends heavily on skilled technicians who have spent years learning the precise, careful craft of winding copper or aluminum conductor around a transformer core in a way that achieves the exact electrical and mechanical specifications the design requires, with the insulation properly placed and the windings properly braced to withstand decades of operation and the occasional severe short-circuit event described in Chapter 12.

This skilled workforce has, in many manufacturing facilities, aged alongside the transformer fleet itself, with a significant fraction of experienced winders and other specialized production staff approaching retirement age, and manufacturers reporting genuine difficulty recruiting and training replacement workers quickly enough to offset both attrition and the production capacity expansion the current demand surge would otherwise justify. Training a new transformer winder to the level of skill and reliability the largest, most demanding designs require is not a matter of weeks; industry sources commonly describe an apprenticeship-style training period measured in years before a new worker can be trusted with the most critical, highest-voltage winding work independently.

Testing Capacity as a Further Constraint

Even after a transformer is manufactured, it must undergo the extensive testing regime described in Chapter 12 before it can be shipped and installed, and high-voltage testing facilities capable of subjecting the largest, highest-voltage transformers to the full battery of required tests are themselves a limited, specialized resource, sometimes shared across multiple manufacturing facilities or even multiple companies, adding a further, less-discussed constraint to the overall time required to bring a newly ordered transformer from initial order to field-ready delivery.

A Genuinely Structural Bottleneck

Taken together, these constraints – specialized steel production requiring years to expand, globally traded copper facing competing electrification-driven demand, specialized insulating materials, and a skilled labor force that cannot be trained quickly regardless of available capital – explain why the transformer bottleneck examined throughout this book's final section is not, in any meaningful sense, a problem that can be solved simply by utilities offering manufacturers more money or placing larger orders. Each of these constraints is fundamentally a matter of physical manufacturing capacity and human skill development, both of which take years to expand meaningfully, regardless of how much financial incentive exists to accelerate the process.

This does not mean nothing can be done. The next chapter examines the substantial efforts already underway – new domestic manufacturing investment, federal policy responses, and international supply chain diversification – aimed at expanding transformer manufacturing capacity over the coming years, and offers an honest assessment of how much these efforts are likely to help, and on what realistic timeline.

The Global Concentration of Electrical Steel Production

It is worth being specific about just how concentrated grain-oriented electrical steel production actually is, because the figure surprises many people encountering it for the first time. Global production capacity for this specialized material is concentrated among a relatively small number of major steel producers, including facilities in the United States, Japan, South Korea, China, and a handful of European countries, with the largest single producers each representing a substantial share of total global output. This concentration means that any disruption affecting even one or two major producers – whether from a plant outage, a trade dispute, or a decision to prioritize a particular geographic market – can meaningfully affect global availability and pricing for the entire downstream transformer manufacturing industry, a vulnerability with few close parallels in most other major industrial supply chains, where production capacity is typically spread across a larger number of more geographically diverse facilities.

Trade Policy Complications

Trade policy has added a further layer of complexity to the electrical steel and transformer supply chain in recent years, with tariffs imposed on imported steel products, including grain-oriented electrical steel, under various trade actions adopted by the United States over the past decade, intended primarily to protect domestic steel producers from foreign competition but carrying the secondary effect of raising input costs for domestic transformer manufacturers who rely on imported electrical steel to supplement limited domestic production capacity. This tension – between trade policy goals focused on protecting domestic steel production and grid-reliability goals focused on ensuring transformer manufacturers have affordable access to the specialized steel their products require – has become an active subject of policy debate among manufacturers, utilities, and federal officials working to address the broader transformer bottleneck described throughout this book's final section.

Apprenticeship Programs and Workforce Development

In response to the skilled labor constraints described in this chapter, several major transformer manufacturers, along with industry trade associations and, in some cases, state workforce development agencies, have expanded apprenticeship and technical training programs specifically aimed at developing the next generation of transformer winders, testers, and other specialized production staff. These programs typically combine structured classroom instruction in electrical theory and manufacturing processes with extensive hands-on, supervised production experience, reflecting the genuinely craft-based nature of transformer manufacturing described earlier in this chapter. While these programs represent a genuine and necessary long-term response to the workforce constraint, industry representatives have generally been candid that meaningfully expanding the specialized skilled workforce will take years to accomplish, tracking closely with the multi-year manufacturing capacity expansion timelines discussed in the following chapter, rather than offering any prospect of near-term relief to the current bottleneck.

Comparing the Transformer Supply Chain to Semiconductors

Industry analysts and journalists covering the transformer bottleneck have increasingly drawn a comparison to the semiconductor shortage that disrupted automotive and electronics manufacturing during the early 2020s, and the comparison is instructive both for its similarities and its differences. Like advanced semiconductor manufacturing, large power transformer production depends on a small number of highly specialized facilities, substantial capital investment to expand capacity, and materials or processes that cannot be easily substituted or sourced from a broader range of alternative suppliers on short notice. Unlike semiconductors, however, transformer manufacturing depends less on the kind of extraordinarily precise, cutting-edge process technology that limits semiconductor fabrication to a tiny handful of companies worldwide, and more on a combination of specialized but comparatively more broadly available materials science, substantial capital equipment, and, critically, the kind of hands-on skilled labor discussed in this chapter – meaning the transformer bottleneck is, in principle, somewhat more addressable through capacity expansion and workforce development than the semiconductor shortage was, even though both share the fundamental characteristic of requiring years, not months, to meaningfully resolve.

Why New Entrants Have Struggled

The favorable pricing environment described in Chapter 17 has attracted a number of new entrants seeking to establish transformer manufacturing operations, including some backed by private equity investment specifically targeting the sector's current supply-demand imbalance. Several of these new entrants have encountered the same fundamental constraints described throughout this chapter: difficulty securing reliable access to grain-oriented electrical steel, given existing long-term supply relationships between established manufacturers and steel producers; difficulty recruiting sufficiently experienced engineering and production staff, given the industry-wide skilled labor shortage; and the multi-year timeline required to establish the testing capability and quality-assurance track record utility customers require before trusting a new manufacturer with orders for critical, expensive, long-lead-time equipment. None of these obstacles is necessarily insurmountable, but together they illustrate why the current favorable pricing environment has not, by itself, been sufficient to produce a rapid expansion of new manufacturing capacity from new market entrants, reinforcing this book's broader argument that the transformer bottleneck reflects genuine physical and human-capital constraints rather than a simple failure of market incentive.

Quality Control and the Cost of Cutting Corners

Given the intense pressure manufacturers face to increase output amid the current demand surge, it is worth addressing directly a concern industry observers have raised: whether the pressure to accelerate production might tempt some manufacturers, particularly newer entrants eager to capture market share described earlier in this book's final section, to cut corners on the quality control and testing processes described in Chapter 12. Established manufacturers with long track records and reputations built over decades generally have strong institutional incentives to resist this temptation, since a single high-profile transformer failure traceable to inadequate quality control could cause lasting reputational damage far exceeding any short-term production gains. Utilities, for their part, have generally maintained rigorous acceptance testing requirements even amid the current supply pressure, unwilling to accept equipment that has not been properly validated regardless of how urgently they need additional capacity.

Nonetheless, industry analysts and some utility engineers have expressed concern about newer or less established manufacturers, particularly those manufacturing in regions with less mature regulatory oversight of industrial quality standards, and several documented cases of transformer quality problems traced to specific overseas manufacturing facilities have prompted some utilities to tighten their manufacturer qualification and inspection requirements even as they simultaneously seek to diversify their supplier base, described in Chapter 19, to reduce dependence on any single manufacturer or country. This tension – between the urgent need for additional manufacturing capacity and the equally urgent need to maintain the rigorous quality standards this book's Part Three has described as essential to long-term reliable operation – represents one more genuine complexity within the broader transformer bottleneck story, and one utilities and regulators will likely continue navigating carefully as new manufacturing capacity, from both established and newer producers, comes online over the remainder of the decade.

Chapter 19: The Race to Build More

Faced with the demand surge, aging fleet, and structural manufacturing constraints described in the previous four chapters, utilities, manufacturers, policymakers, and even some of the technology companies driving AI datacenter demand have begun mounting a genuine, if necessarily gradual, response. This chapter examines the most significant of these efforts, and offers a candid assessment of their likely impact and realistic timeline.

New Domestic Manufacturing Investment

Several major transformer manufacturers, along with a number of newer entrants seeking to capitalize on the current demand surge and favorable pricing environment described in Chapter 17, have announced new domestic manufacturing capacity investments over the first half of the 2020s, including new factory construction, expansions of existing facilities, and, in some cases, the reopening or repurposing of previously idled industrial sites. These investments, welcome as they are, illustrate directly the structural timeline problem described throughout Part Four: constructing, equipping, staffing, and fully ramping up a new large power transformer manufacturing facility, or even substantially expanding an existing one, is itself a multi-year undertaking, typically requiring somewhere in the range of two to four years from initial investment decision to meaningful production output, even before accounting for the additional time required to train the skilled workforce discussed in Chapter 18.

This means that even the most aggressive, well-funded manufacturing capacity expansion announcements made in the early-to-mid 2020s are not expected to deliver substantial additional transformer supply until the latter half of the decade at the earliest, and the full benefit of new grain-oriented electrical steel production capacity, discussed in the previous chapter, may not be felt until even later, given that steel production capacity expansion generally must precede, or at minimum proceed in careful coordination with, transformer manufacturing capacity expansion. Utilities and industry analysts tracking this buildout generally describe the transformer supply situation as likely to remain tight through the remainder of the 2020s, with meaningful easing, if current investment trends continue, becoming more plausible toward the end of the decade and into the early 2030s.

Federal Policy Responses

The federal government has responded to the transformer bottleneck through several distinct policy channels. The Department of Energy has used authority under the Defense Production Act to designate certain grid-critical equipment, including large power transformers and grain-oriented electrical steel, as strategic priorities eligible for expedited government support and, in some cases, direct financial assistance to encourage domestic manufacturing capacity expansion. Federal infrastructure and industrial policy legislation adopted in the early 2020s has similarly directed funding toward grid resilience and domestic manufacturing capacity, including provisions specifically aimed at expanding the domestic transformer and grid-component supply chain, in recognition of both the reliability concerns discussed throughout this book and broader economic and national security concerns about dependence on a small number of overseas manufacturers for equipment as critical as large power transformers.

Regulatory bodies, including state utility commissions and the Federal Energy Regulatory Commission, have also begun adjusting some interconnection and procurement processes in response to the transformer bottleneck, including efforts to allow utilities greater flexibility to order equipment further in advance of confirmed need, to standardize transformer specifications across utilities in ways that could allow manufacturers to build more units to common designs rather than fully bespoke specifications for every order, and to explore whether some interconnection customers, including datacenter developers, might be permitted or encouraged to provide their own equipment or financing to accelerate specific projects, an approach examined further below.

Datacenter Developers Taking Matters Into Their Own Hands

Perhaps the most striking market response to the transformer bottleneck, and one *The Reliability Frontier* touched on from the demand side without examining the underlying equipment dynamics this book has now provided, has been the willingness of some hyperscale technology companies developing AI datacenters to directly fund, and in some cases directly procure, the transformer and transmission equipment needed to serve their own projects, rather than waiting for a utility's normal procurement and rate-recovery process to run its course. This can take several forms: direct financial contributions toward utility infrastructure upgrades in exchange for expedited interconnection timelines, direct equipment procurement arrangements in which the datacenter developer effectively places its own order with a transformer manufacturer and coordinates delivery with the serving utility, and, in a small but growing number of cases, on-site or co-located generation arrangements – including natural gas generation, and increasingly proposals involving small modular nuclear reactors, discussed further below – that reduce, though rarely eliminate entirely, the developer's dependence on new transmission-connected transformer capacity.

These developer-driven strategies illustrate a broader theme worth naming directly: the companies with the greatest financial resources and the most urgent timelines have some ability to work around, or at least accelerate their position within, the transformer bottleneck, by essentially paying a premium to jump the manufacturing queue or by developing alternative supply arrangements unavailable to smaller, less well-capitalized customers. This dynamic raises legitimate equity and grid-planning concerns that state regulators and utility executives are actively grappling with: if the largest, wealthiest datacenter developers can effectively purchase their way to the front of a limited manufacturing queue, what does that mean for the interconnection timelines facing smaller commercial customers, new renewable generation projects, or utilities' own routine equipment replacement needs, all competing for the same constrained supply of large power transformers?

International Supply Chain Diversification

Alongside domestic manufacturing expansion, utilities and equipment brokers have increasingly sought to diversify transformer procurement across a broader range of international suppliers, including manufacturers in South Korea, India, and several European countries, partly to access additional manufacturing capacity beyond the historically dominant American, European, and East Asian producers, and partly to reduce dependence on any single country's or company's production capacity and the associated geopolitical and trade-policy risk that dependence entails – a concern sharpened by the broader geopolitical tensions of the mid-2020s, including trade friction with China, which has itself emerged as an increasingly significant, if geopolitically fraught, source of transformer manufacturing capacity given its own massive, state-supported grid buildout, a comparison *The Reliability Frontier* examined in detail when assessing China's electricity generation capacity additions relative to those of the United States.

A Genuine but Gradual Response

Taken together, the manufacturing investment, federal policy support, developer self-help strategies, and international supply chain diversification described in this chapter represent a genuine, multi-pronged response to the transformer bottleneck, and there is reasonable cause for cautious optimism that the acute crisis conditions of the early-to-mid 2020s will gradually ease over the remainder of the decade, as new manufacturing capacity comes online and the industry adjusts to a new, more elevated baseline of demand. But nothing in this chapter should be read as suggesting the bottleneck will resolve quickly, or that the multi-year lead times, elevated prices, and interconnection delays described in Chapter 17 will disappear on any timeline shorter than several years, given the fundamental physical and human-capital constraints on manufacturing capacity expansion described in Chapter 18.

The final chapter of this book turns from the near-term crisis response examined here to a longer-term, more speculative question: what might the transformer of the future actually look like, as engineers pursue new technologies – including solid-state power electronics and increasingly sophisticated digital monitoring – that could, over a longer horizon, change not just how quickly transformers can be manufactured, but what a transformer fundamentally is and does within an increasingly complex, increasingly renewable, increasingly digitally managed electrical grid.

Standardization as an Underappreciated Lever

One policy and industry response deserving fuller attention is the push toward greater standardization of transformer specifications across utilities, an effort that addresses a subtler but real contributor to the current bottleneck: because large power transformers have historically been engineered to each utility's own bespoke specifications, as described in Chapter 12, manufacturers have had limited ability to build ahead of confirmed orders or maintain meaningful finished-goods inventory, since a transformer built to one utility's specifications may not suit another utility's needs without modification. Industry groups and some federal program offices have promoted the development of a smaller number of standardized reference designs, covering the most common voltage classes and power ratings utilities typically require, that manufacturers could potentially build in advance of confirmed orders, reducing effective lead times for utilities willing to accept a standardized design rather than a fully bespoke one tailored to their exact specifications.

This standardization push faces real obstacles, since utilities' historical preference for bespoke specifications often reflects genuine site-specific engineering requirements, discussed in Chapter 12, rather than mere institutional preference, and persuading a sufficient number of utilities to accept a smaller menu of standardized designs requires overcoming both genuine engineering considerations and understandable institutional caution about deviating from established procurement practices. Nonetheless, several industry initiatives launched in the mid-2020s have made real, if still limited, progress toward this kind of standardization, and analysts tracking the transformer market generally view it as one of the more promising near-to-medium-term levers available for reducing effective delivery timelines without requiring entirely new manufacturing capacity.

Small Modular Reactors and Datacenter Co-Location

The growing interest among AI datacenter developers in small modular nuclear reactors, mentioned briefly above, deserves additional context given its direct relevance to the transformer bottleneck. Small modular reactors, still in relatively early stages of commercial deployment as of the mid-2020s, offer datacenter developers the prospect of dedicated, carbon-free, highly reliable generation capacity co-located with or very near their datacenter facilities, potentially reducing dependence on new transmission-connected transformer capacity of the kind examined throughout this book's final section. Several major technology companies have announced power purchase agreements or direct investment in small modular reactor projects specifically framed around meeting datacenter electricity needs, though the realistic timeline for these projects to reach commercial operation at meaningful scale generally extends well into the late 2020s and early 2030s, meaning small modular reactors are unlikely to meaningfully ease the near-term transformer bottleneck described throughout this chapter, even as they may play an increasingly important role in the somewhat longer-term outlook *The Reliability Frontier* examines in its own discussion of next-generation nuclear technology.

The Equity Question in Fuller Detail

The equity concerns raised briefly earlier in this chapter deserve fuller treatment, because they touch on a genuine tension in how a scarce, critical resource should be allocated across competing uses during a period of extended shortage. When a well-capitalized datacenter developer can offer to directly fund infrastructure upgrades or provide its own equipment financing to secure an expedited interconnection timeline, it is, in effect, using its financial resources to move ahead of other customers – smaller businesses, new housing developments, community solar projects, or the utility's own routine equipment replacement needs – who lack comparable resources to offer similar accelerating payments. State regulators overseeing utility interconnection practices have begun grappling directly with this tension, in some cases developing formal policies governing when and how utilities may accept accelerated-timeline payments from large customers, and in some cases requiring utilities to demonstrate that such arrangements do not unreasonably disadvantage other customers waiting in the same interconnection queue.

This is, in a meaningful sense, a new and still-unsettled area of utility regulatory policy, since traditional interconnection queue management practices, developed over decades of comparatively predictable, gradually growing demand, did not anticipate a scenario in which a single customer's request could represent as much load as an entire utility's normal annual growth, nor a scenario in which that customer might offer to pay a premium specifically to jump ahead of other equally valid, if smaller, requests already in the queue. How this tension resolves over the coming years – through new regulatory frameworks, through expanded manufacturing capacity that eventually reduces the scarcity driving the tension in the first place, or through some combination of both – remains one of the genuinely open questions this book cannot yet answer with confidence, though it is one worth watching closely for any reader following the broader story of how America's grid adapts to the AI era.

Grid Enhancing Technologies as a Complementary Strategy

Beyond the manufacturing-focused responses described in this chapter, utilities have also increasingly deployed a category of technology sometimes called grid-enhancing technologies – including dynamic line rating systems that allow existing transmission lines to safely carry more power under favorable weather conditions, and advanced power flow control devices that can redirect electricity more efficiently across an existing transmission network – as a way to extract more usable capacity from the grid's existing infrastructure, including existing transformers, without necessarily requiring new equipment. While these technologies cannot substitute entirely for new transformer capacity where genuinely new load is being added, they represent a valuable complementary strategy for stretching existing capacity further while new manufacturing capacity, discussed throughout this chapter, gradually comes online over the remainder of the decade.

International Cooperation and Allied Supply Chains

Beyond the domestic manufacturing investment and international diversification discussed earlier in this chapter, the United States has also pursued deeper coordination with allied nations on critical grid equipment supply chains, recognizing that no single country, including the United States, is likely to achieve full domestic self-sufficiency in large power transformer manufacturing within any realistic near-term timeline, given the multi-year capacity expansion timelines discussed throughout this chapter. Coordination with allied producers in countries including South Korea, Japan, and several European nations has focused on ensuring diversified, resilient supply access, while trade and industrial policy discussions have grappled with balancing this need for diversified allied supply against the domestic manufacturing revitalization goals described earlier in this chapter – goals that, if pursued too aggressively through protectionist trade measures, could raise costs and further complicate the very supply chain diversification strategy meant to ease the current bottleneck.

A Realistic Timeline for Recovery

Synthesizing the various response efforts described throughout this chapter, most informed industry observers, utility executives, and policy analysts tracking the transformer market converge on a broadly similar realistic timeline: continued tight supply and elevated pricing through the remainder of the 2020s, with gradual, meaningful easing becoming plausible toward the very end of the decade and into the early 2030s, as new domestic and allied manufacturing capacity, new grain-oriented electrical steel production capacity, and an expanded, better-trained skilled workforce all come online in a coordinated, mutually reinforcing way. This timeline is not a certainty – it depends on sustained investment, continued policy support, and the demand side of the equation, particularly the pace of AI datacenter development, not accelerating faster than the supply side can plausibly follow – but it represents the most credible, evidence-based outlook available as of this book's writing, and it is offered here with the appropriate humility that any forecast about a genuinely fast-moving, still-unfolding industrial and technological story deserves.

Chapter 20: The Transformer of the Future

This book has told the story of a single machine across nearly two centuries: Michael Faraday's iron ring in 1831; the Hungarian engineers who made it commercially practical in 1885; William Stanley's Great Barrington demonstration and the War of the Currents it helped decide; the giant units that carried Niagara Falls' power to Buffalo and, eventually, that built out an entire interconnected national grid; the humble pole-mounted transformer that finally brought electricity to the American farmhouse; and the aging, strained fleet that now sits, quite literally, at the center of whether America's grid can meet the surging demand of the AI era chronicled throughout Part Four. This final chapter looks forward, examining the emerging technologies that may reshape what a transformer is and does over the coming decades, while remaining honest about how much of that future remains genuinely uncertain.

Solid-State Transformers

The most technologically ambitious development on the horizon is the solid-state transformer, a device that uses power electronic components –

semiconductor switches similar in underlying principle, though very different in application, to those found in computer processors and power supplies – to convert voltage using high-frequency electronic switching rather than the purely magnetic, Faraday-principle induction that has defined every transformer discussed in this book so far. A solid-state transformer can, in principle, offer capabilities no conventional transformer can match: the ability to actively control power flow and voltage in real time, rather than simply presenting a fixed voltage-transformation ratio; the ability to interface directly and efficiently with the direct-current output of solar panels and battery storage systems, an increasingly important capability given the renewable energy integration challenges *The Reliability Frontier* examined at length; and, in some designs, a substantially smaller physical footprint than a conventional transformer of equivalent power rating.

These capabilities have made solid-state transformers a subject of intense research and, increasingly, early-stage commercial pilot deployment throughout the mid-2020s, particularly for applications involving renewable energy integration, electric vehicle fast-charging infrastructure, and grid-edge applications where the ability to actively manage power flow offers particular value. It would be premature, however, to suggest solid-state transformers are poised to replace conventional transformers at scale in the near future, for reasons directly related to the manufacturing and materials themes running throughout this book's final section: solid-state transformers require substantial quantities of power semiconductor components, an industry facing its own well-documented supply chain and manufacturing capacity constraints, and the technology remains, as of the mid-2020s, generally more expensive per unit of power-handling capacity than conventional transformer technology, particularly at the very largest, extra-high-voltage scale where conventional transformers remain most dominant and most difficult to displace. Most industry analysts and grid engineers expect solid-state transformers to find their earliest and most significant adoption in specific, high-value niche applications – renewable energy interconnection, EV fast-charging, and certain grid-edge distribution applications – well before, if ever, displacing conventional oil-filled transformers at the scale of the giant units discussed in Chapter 12.

Grid-Forming Technology and the Changing Character of Grid Stability

A related and increasingly important development, discussed extensively in *The Reliability Frontier*'s treatment of renewable energy integration, involves so-called grid-forming inverters and, by extension, grid-forming transformer and power-electronics systems more broadly – technology designed to help a grid increasingly composed of inverter-based renewable and battery resources maintain the frequency and voltage stability that traditional, large rotating generators historically provided as an inherent physical byproduct of their mechanical operation. As the American generation mix continues shifting toward solar, wind, and battery storage, and away from the large synchronous generators that have historically anchored grid stability, transformers and the power electronics increasingly associated with them will need to take on some of this stabilizing role more actively than the purely passive, voltage-transforming function this book has described throughout its historical chapters. This represents a genuinely significant evolution in the transformer's functional role within the grid, even as the underlying Faraday-principle physics described in Chapter 1 continues to govern the vast majority of transformers actually in service.

Digital Monitoring and Predictive Maintenance

A nearer-term and, in some ways, more immediately consequential development involves the increasing integration of digital sensors and monitoring systems into both new and existing transformers, building on the dissolved gas analysis and condition-monitoring practices described in Chapter 11. Modern sensor packages can continuously track a transformer's internal temperature, oil condition, partial discharge activity, and other indicators of developing problems, feeding this data to utility control centers and increasingly to machine-learning-based analytics systems designed to flag developing equipment problems earlier and more reliably than periodic manual inspection alone can achieve.

This kind of predictive maintenance capability offers real value given the aging-fleet vulnerability described in Chapter 15: rather than relying solely on scheduled inspection intervals or waiting for outright failure, utilities can increasingly identify specific transformers showing early signs of developing problems and prioritize those units for closer monitoring, targeted maintenance, or replacement, extracting more useful, safely managed service life from an aging fleet while better targeting the limited new-transformer manufacturing capacity described throughout this book's final chapters toward the units that most urgently need replacement, rather than replacing units on a rigid, age-based schedule that may not accurately reflect their actual remaining useful life.

A Sober Assessment

It would be satisfying to end this book with a confident prediction that emerging technology will soon resolve the bottleneck described throughout Part Four. Intellectual honesty requires resisting that temptation. Solid-state transformers, grid-forming technology, and digital monitoring all represent genuine, valuable engineering progress, and each will likely play an increasingly important role in the American grid over the coming decade. None of them, however, addresses the core physical manufacturing capacity constraints – grain-oriented electrical steel production, skilled winding labor, and high-voltage testing capacity – described in Chapter 18, which remain the binding constraint on how quickly the conventional, Faraday-principle transformers that still carry the overwhelming majority of American electricity can be built. The transformer's future, in the near term at least, looks less like a technological leap and more like the patient, capital-intensive work of building more factories, training more skilled workers, and expanding the supply of specialized steel – unglamorous work, but precisely the kind of work this book has shown, again and again, actually built and sustained the American electrical grid across nearly a century and a half.

Conclusion: The Machine That Will Decide

This book began with an image: electricity traced backward from your kitchen outlet, through a small gray canister on a utility pole, through a substation, along a transmission line, to the giant transformers standing beside a power plant, each one silently, invisibly doing the same essential job Michael Faraday first demonstrated with an iron ring in a London laboratory in 1831. It closes with a different but related image: an interconnection queue, somewhere in America, where a utility engineer is explaining to a datacenter developer, a solar project developer, or simply a growing local manufacturer, that the equipment needed to serve their new electrical load will not arrive for another three or four years, because the factories that build it cannot go faster.

Both images describe the same machine. That continuity is the central argument of this book. The transformer that decided the War of the Currents, that carried Niagara Falls' power to Buffalo, that finally reached the American farmhouse in the 1930s and 1940s, and that now stands as the primary bottleneck constraining how quickly America's grid can meet the demands of artificial intelligence, electrified transportation, reshored manufacturing, and a warming climate, is not a metaphor or an abstraction. It is a specific, physical, remarkably unchanged piece of engineering, still built around the same core-and-winding principle Faraday demonstrated nearly two centuries ago, and its availability – or scarcity – will help determine, as concretely as any single factor can, how quickly the United States can build the electrified, AI-powered, increasingly renewable economy so much current policy and investment assumes is coming.

My earlier book, *A History of the United States Electrical Grid*, told the story of the institutions, regulations, and markets that built and governed that grid. My more recent book, *The Reliability Frontier*, told the story of the reliability strain that grid now faces as artificial intelligence and electrification reshape American electricity demand. This book has told the story of the machine sitting quietly underneath both of those larger narratives, doing the essential, invisible work that made every chapter of both stories possible – and that now, after a century of being taken for granted, has become, unexpectedly, one of the most consequential pieces of industrial equipment in America.

The gray canister on the pole outside your window is not going to look any different tomorrow than it did yesterday. But I hope, having read this book, you will look at it differently than you did before.

High-Temperature Superconductors: A More Distant Possibility

Among the more speculative technologies sometimes discussed in connection with the transformer's future is the potential application of high-temperature superconducting materials to transformer winding design, which could, in principle, dramatically reduce the resistive losses that ordinary copper or aluminum windings inevitably produce, and could potentially allow more compact transformer designs for a given power rating. Superconducting transformer prototypes have been built and tested in limited research and demonstration contexts, but the practical challenges of maintaining the extremely low temperatures most superconducting materials require for large-scale, continuously operating grid equipment remain substantial, and most industry observers view large-scale commercial superconducting transformer deployment as a considerably more distant prospect than the solid-state and digital-monitoring technologies discussed earlier in this chapter, likely, if it happens at all, unfolding over a timeline measured in decades rather than years.

Recycling and the Circular Economy for Transformers

A final, more near-term development worth noting involves growing industry attention to transformer recycling and material recovery, as aging units are retired from service. A retired transformer contains substantial quantities of recoverable copper or aluminum, along with grain-oriented electrical steel that, while not directly reusable in a new transformer core given the precision manufacturing the material requires, can be recycled into other steel products, and the mineral oil, if not contaminated, can in some cases be reprocessed and reused. Given the raw material constraints discussed in Chapter 18, several manufacturers and recycling firms have expanded efforts to recover and reprocess these materials more systematically from the wave of aging transformers reaching the end of their service life, a modest but genuine contribution to easing the raw material pressure described throughout this book's final section, even if recycled material alone cannot come close to meeting the industry's overall raw material needs given the scale of new manufacturing demand described in Chapter 16.

Closing Thought on the Machine's Continuity

Every technology discussed in this chapter – solid-state power electronics, grid-forming control systems, digital condition monitoring, superconducting materials, and improved recycling practices – represents a genuine evolution of the transformer concept, but none of them replaces the fundamental Faraday-principle insight with which this book began: a changing magnetic field, linking two windings through a shared magnetic core, can move electrical energy from one circuit to another while changing its voltage along the way. Nearly two centuries after Faraday wound two coils around an iron ring in a London laboratory, that insight remains the foundation of the machine quietly determining how quickly America can build the electrified future so much current policy, investment, and public conversation assumes is already underway.

What Readers Should Watch For in the Years Ahead

For readers who want to track how this story continues to unfold beyond this book's publication, several specific indicators are worth watching. First, transformer lead times themselves, which utilities and equipment brokers report regularly and which trade publications track closely; a sustained decline from the three-to-five-year figures described in Chapter 17 would be the clearest sign that the new manufacturing capacity discussed in Chapter 19 is genuinely beginning to ease the bottleneck. Second, grain-oriented electrical steel production capacity announcements and, more importantly, their actual completion and ramp-up, since this specific material constraint, discussed in Chapter 18, sits further upstream in the supply chain than transformer assembly itself and therefore serves as something of a leading indicator for the broader industry's capacity to expand. Third, the pace and outcome of interconnection queue reform efforts at the state and federal level, which will determine how efficiently the transformer capacity that does exist gets allocated across competing demands during the continued period of scarcity this book's final chapters describe. And fourth, the actual build-out pace of AI datacenter capacity itself, since continued acceleration in announced projects would sustain or worsen current pressures, while any moderation in the pace of new datacenter announcements – whether due to shifts in AI investment economics, improved chip efficiency, or the emergence of viable co-located generation alternatives discussed in Chapter 19 – could meaningfully ease the demand side of the equation this book has described.

A Final Word on Humility

This book has made a sustained argument that the transformer bottleneck reflects genuine, difficult-to-resolve structural constraints, and that argument is offered with real confidence given the weight of evidence this book's final section has presented. But readers should hold any specific prediction about exactly when or how thoroughly this bottleneck resolves with appropriate humility. Industrial supply chains, technology adoption curves, and policy responses all contain genuine uncertainty, and the specific pace of AI datacenter development, in particular, has already surprised most forecasters, in both directions, at various points over the past several years. What this book can say with confidence is the underlying mechanism: transformers take years to manufacture because of genuine physical and human-capital constraints, those constraints are being addressed through real, substantial investment, and the resulting easing, when it comes, will arrive gradually rather than suddenly, over a period of years rather than months. Beyond that, the precise timeline remains, appropriately, an open question – one this book's readers are now considerably better equipped to follow and evaluate than they were before picking it up.

Workforce Development as the Longest Pole in the Tent

Of all the emerging responses discussed throughout this book's final section, this closing chapter returns one last time to the theme that most consistently limits how quickly any of them can meaningfully change the picture: skilled human labor. Solid-state transformers, grid-forming control systems, and digital monitoring platforms all still require engineers and technicians trained to design, install, maintain, and troubleshoot them, and the same workforce constraints discussed in Chapter 18 apply, in modified form, to these newer technologies as much as to conventional transformer manufacturing. Community colleges, technical schools, and utility-sponsored training programs across the country have expanded electrical trades and power engineering curricula in response to the demand this book has documented throughout its final chapters, and industry observers broadly agree that sustained investment in this kind of technical workforce development represents one of the highest-leverage, if least glamorous, responses available to policymakers and industry leaders seeking to ease the transformer bottleneck over the coming decade.

A Final Return to the Machine Itself

This book has traveled a long distance, from a London laboratory in 1831 to a datacenter interconnection queue in the mid-2020s, but it has never truly left the single machine at its center. The transformer that Faraday's ring made conceptually possible, that the Ganz Works engineers made commercially practical, that William Stanley proved could work on an American Main Street, that carried Niagara Falls' power to Buffalo, that finally reached the American farmhouse, and that now determines how quickly a datacenter, a battery factory, or a growing suburb can get the electricity it needs, has remained, at its physical core, remarkably consistent across nearly two centuries. What has changed, again and again, is the scale of ambition placed upon it, and the industrial capacity available to meet that ambition. The story of the twenty-first century's electrical grid, like the story of every century before it, will be written substantially in transformer factories – and readers who have followed this book to its end now understand, in a way few of their neighbors do, exactly why that quiet, unglamorous fact matters as much as it does.

BACK MATTER

Glossary of Key Terms

Alternating current (AC). Electric current that periodically reverses direction, oscillating sixty times per second in the standard American grid. Its continuously changing nature is what allows transformers to function, since only a changing magnetic field induces current in a neighboring winding.

Bushing. A heavily insulated fitting that allows a high-voltage electrical connection to pass safely through a transformer's grounded steel tank wall.

Core. The laminated iron or electrical-steel structure at the heart of a transformer, which concentrates and channels the magnetic field linking the primary and secondary windings.

Core-form / shell-form. Two broad transformer core geometries: in a core-form design, windings surround the core's legs; in a shell-form design, the core surrounds the windings on multiple sides.

Direct current (DC). Electric current that flows steadily in one direction. Because a steady current produces an unchanging magnetic field, direct current cannot be efficiently transformed to a different voltage using the induction principle that powers AC transformers.

Dissolved gas analysis (DGA). A diagnostic technique that measures gases dissolved in transformer oil to detect early signs of internal overheating or electrical arcing before outright failure occurs.

Distribution transformer. A smaller transformer, often pole- or pad-mounted, that steps voltage down from the distribution-system level (typically a few thousand to around 13,000 volts) to the 120/240 volts used by homes and small businesses.

Eddy currents. Unwanted circulating currents induced within a solid conductor, such as a transformer core, by a changing magnetic field; they waste energy as heat and are minimized through laminated core construction.

Electromagnetic induction. The physical principle, discovered by Michael Faraday in 1831, whereby a changing magnetic field induces an electric current in a nearby conductor. It is the foundational principle behind every transformer.

Grain-oriented electrical steel. A specialized steel alloy, processed to align its internal crystal structure, used in transformer cores because it channels magnetic fields with unusually low energy loss.

Grid-forming inverter. A power-electronics device, increasingly important as renewable and battery resources replace traditional generators, that can help establish and maintain grid voltage and frequency rather than simply following a signal set by other equipment.

Impedance. A transformer's internal resistance to current flow, which affects fault-current behavior on the connected system and must be engineered to coordinate with a utility's protective relay settings at a specific installation site.

Interconnection queue. The formal process, and the backlog of pending requests within it, through which new generation projects or large electricity customers apply to connect to a utility's transmission or distribution system.

Lamination. The construction technique of building a transformer core from thin, individually insulated sheets of steel rather than a solid block, which dramatically reduces eddy-current losses.

Large power transformer (LPT). Industry shorthand for the largest category of transformer, typically used at power plants and major substations, often weighing over 100 tons and custom-engineered to a specific utility's specifications.

Load tap changer. A mechanical device allowing a transformer's turns ratio, and therefore its output voltage, to be adjusted incrementally while the unit remains energized and in service.

Mobile substation. A trailer-mounted transformer and associated switching equipment that can be deployed temporarily to restore partial service after an unexpected transformer failure, pending a permanent replacement.

Parallel connection. An electrical arrangement, pioneered commercially by the ZBD transformer in 1885, in which multiple customer loads draw independently from a shared voltage source, allowing one customer's usage to be switched on or off without affecting others.

Primary winding / secondary winding. The input and output coils of a transformer, respectively; the ratio of turns between them determines the transformer's voltage transformation.

Schnabel car. A specialized railcar, first developed in the 1930s, engineered so that a transformer's own tank becomes part of the load-bearing structure during rail transport, distributing its enormous weight across additional axles.

Solid-state transformer. An emerging transformer technology that uses high-frequency semiconductor switching, rather than purely magnetic induction, to transform voltage, offering active power-flow control not available from conventional designs.

Transformer oil. A specially formulated mineral oil (or, increasingly, synthetic ester fluid) used inside large transformers as both an electrical insulator and a cooling medium.

Transmission voltage. The high voltage – commonly 138,000 to 765,000 volts in the United States – at which electricity travels over long distances between generating plants and regional substations, minimizing resistive losses.

Turns ratio. The ratio between the number of winding turns on a transformer's primary and secondary coils, which directly determines its voltage transformation ratio.

A Note on Sources and Further Reading

This book draws on a wide range of historical and contemporary sources, and readers interested in going deeper on any particular thread will find rich material available. For the nineteenth-century invention history covered in Part One, the surviving technical papers and patents of Michael Faraday, Ottó Bláthy, Miksa Déri, Károly Zipernowsky, and William Stanley remain available through academic archives and, in many cases, digitized patent records, alongside numerous biographies of Thomas Edison, George Westinghouse, and Nikola Tesla that cover the War of the Currents from various angles, some considerably more sensationalized than the account offered in Chapter 5 of this book.

For the grid-expansion history covered in Part Two, readers should consult my earlier book, *A History of the United States Electrical Grid*, which examines the regulatory, financial, and institutional dimensions of this same period in considerably greater depth than this book's transformer-focused account attempts. Historical accounts of the Rural Electrification Administration and the broader rural electrification movement, including oral histories collected from farm families who lived through the transition described in Chapter 9, offer a valuable and often moving complement to the engineering-focused account given here.

For the technical material in Part Three, readers with a stronger technical background may wish to consult formal transformer engineering references published by the Institute of Electrical and Electronics Engineers and by major equipment manufacturers, which provide considerably more mathematical and design detail than this book's general-reader treatment attempts or requires.

For the contemporary supply-chain crisis examined in Part Four, the most current information is found not in books, which inevitably lag fast-moving industry developments, but in ongoing reporting from energy-industry trade publications, periodic assessments published by the U.S. Department of Energy on large power transformer reliability and supply chain risk, and the public regulatory filings utilities make before state utility commissions and the Federal Energy Regulatory Commission, which contain detailed, if technical, testimony on current market conditions. Readers interested in the demand side of this story – the AI datacenter boom, renewable energy integration, and the broader reliability challenges facing the American grid in the 2020s – should consult my companion volume, *The Reliability Frontier: How America's Grid Is Racing to Meet the AI Era*, which examines these themes in considerably greater depth from the perspective of grid operations and policy.

Acknowledgments

This book would not exist without the two volumes that preceded it, and I remain grateful to the readers of both *A History of the United States Electrical Grid* and *The Reliability Frontier* whose questions about transformers, asked more often and more insistently than I initially expected, convinced me this machine deserved a book of its own. Any errors of fact or interpretation in the pages above remain, of course, my own.

Douglas Donovan

www.ingramcontent.com/pod-product-compliance
Lightning Source LLC
LaVergne TN
LVHW041208061126
851857LV00018B/1979

* 9 7 8 1 0 8 8 1 5 2 8 5 0 *